THE WORKS OF MATRIX DESIGN

矩阵纵横设计作品精选 II 2010—2015

矩阵纵横设计团队 编著

·广州·

图书在版编目（CIP）数据

矩阵纵横设计作品精选.2，2010—2015：英汉对照/ 矩阵纵横设计团队编著. — 广州：华南理工大学出版社，2015.5

ISBN 978-7-5623-4607-4

Ⅰ．①矩… Ⅱ．①矩… Ⅲ．①室内装饰设计－作品集－中国－现代 Ⅳ．①TU238

中国版本图书馆CIP数据核字（2015）第074077号

矩阵纵横设计作品精选 Ⅱ 2010—2015 The Works of Matrix Design Ⅱ 2010—2015
矩阵纵横设计团队 编著

出 版 人：韩中伟
出版发行：华南理工大学出版社
　　　　　（广州五山华南理工大学17号楼，邮编510640）
　　　　　http://www.scutpress.com.cn　E-mail: scutc13@scut.edu.cn
　　　　　营销部电话：020-87113487　87111048（传真）
策划编辑：赖淑华
责任编辑：李彩霞
印 刷 者：深圳市汇亿丰印刷科技有限公司
开　　本：595 mm × 1020 mm　1/16　　印张：25.5
成品尺寸：248 mm × 290 mm
版　　次：2015年5月第1版　2015年5月第1次印刷
定　　价：398.00元

版权所有　盗版必究　　印装差错　负责调换

matrix®
INTERIOR DESIGN
INTRODUCTION

Matrix Design Co, Ltd, founded in early 2010, is a team of young and prominent designers. Matrix's design philosophy is to create a new world by obeying the rules of the things themselves. Everything in the world has its own disciplines, rules and laws; to follow them is to create a brand new world!

矩阵纵横成立于2010年初,由一批年轻新锐设计师组成,"以矩为方圆,万物皆成阵"是公司的核心理念。世界万物都有各自的规律、规矩、法则,当你去遵循它,将会创造出一个新的世界!

Matrix design © 2010 — 2015

Preface for fifth anniversary
A letter to brothers and sisters

16th March, 2015 is the fifth anniversary day of Matrix. Two years ago we published the first memoir album recording the company's developments. It's not hard to see that it is a green and immature album, but it doesn't matter, as its significance is to record truly what we have been through. Do things in our own way; this is our belief! We become more mature and solid, more confident and steady, and more aware of the responsibilities on shoulder. All these years we treat our works as if we are preparing for wars in our lifetime. I want to say to all my partners: Thank you! No matter you left or stay, nobody can ever deny the difficulties we've overcome and the victories we've achieved together! And all my brothers and sisters who have been all the way persistent in our dream and career, it is an honor for my whole lifetime to work and strive with you! The way you breathe determines the way you survive! Let's fight for the next 5 years hand in hand! Everybody has an ambition, yet mine is not to be worshiped by thousands or followed by hundreds, but is very simple — to make for all my partners a better life, we as a team grow together and enjoy happiness together. Survive and be remembered — this is my original intention. Yes, there are so much I can not do yet, but please trust that I'm trying! Let's fulfill it together! The clarion of our war has just arisen! Trust the power of trust is our belief! In publishing of our second album, I, on behalf of the 3 founders of Matrix, Idmen Liu, Zhaobao Wang and myself, thanks again for all your efforts! Principles in Matrix, Rules in All!

2014.12.20 noon Unmixme Wang

五周年序文 —— 给兄弟姐妹们的一封信

2015年3月16日是公司成立五周年的纪念日，两年之前我们出版了第一本纪念专辑，里面记录了公司成长的点点滴滴。不难从作品中看出稚嫩与青涩，但没关系，其本意就是记录我们携手成长的轨迹。不用管旁人的看法，这是我们战斗的理念！经过岁月的洗礼，我们更加成熟稳健，更加自信沉着，更加清晰自己肩上所负的责任。这些年面对这个战场，是的，我始终以备战的状态面对工作，我想在此对所有的伙伴说声——谢谢！不管是离开还是加入，谁都不能抹去我们曾经共同面对的困难和胜利后的喜悦！还有矢志不渝的兄弟姐妹们，能与你们一起战斗是我今生的荣幸！如何呼吸决定如何生存！让我们携手面对下一个五年吧！说没有野心是假的，但是我的野心不是成为万人瞩目或者随者无数，我的想法很简单，让我们一起战斗的伙伴们吃饱穿暖，一起成长并幸福安康。生存下去并在世上留下印记，这便是我最原始朴素的想法。是的，还有很多我无法做到的，但请相信我在努力！帮我实现它吧！战斗才刚刚打响！崇尚相信的力量就是我们的信仰！在第二本专辑出版之际，我谨代表我自己、刘建辉及王兆宝三位矩阵纵横创始人，对大家所付出的努力表示再次的感恩！以矩为方圆，万物皆成阵！

2014.12.20午 王冠

About the dream
Happy life, calm design

Happy life, calm design is a magnificent vision as well as an easy target. Happiness is abstract, literally meaning the psychological feeling of joy, satisfaction, gratitude and abundance due to the improved human condition. Matrix has, since its establishment in early 2010, kept adjusting and improving its condition, always giving us the happiness, spurring us to change and to go beyond ourselves for both life and design. Under the background of China's rapid development, Chinese design, though still with many problems, has gradually realized the sound development. Benefited from the background, Matrix strides forward without messing up steps; we work hard while intentionally slow down for review and reflection. Design is the commercialization of art; we enjoy seeking a balance between commerce and art. Having a clear goal is precious in today's society. Matrix, after passing through its first 5 years, is advancing toward the next 5 years full of mystery and imagination. Our ultimate goal is Back to Orient, following the call and trend of individuals, teams and societies. Since the door was forced open more than a hundred years ago, China has been kidnapped by the western civilization, resulting in the lack of confidence and firmness after constantly learning, losing and falling. It is our status and inevitable result, as it is unnatural growth, just floating on the surface. Let life, aesthetics and design return to nature, like the oriental wisdom WABI-SABI advocated by Europeans, pursuing simple, quiet, humble, natural life and design. The goal might take years to be proved, but it is clear and firm enough. We have finished over one hundred domestic projects in the 5 years. We couldn't have achieved it without the support and trust from our clients as well as the efforts taken by every member in the team. I myself have been benefited from the team and the two partners, Unmixme Wang and Zhaobao Wang. We promise ourselves that the dream will come true, as is said by Alibaba's Jack Ma, dream is necessary, in case it might be realized :)

2015.01.12 Idmen Liu

关于梦想—— 幸福生活，从容设计

幸福生活，从容设计，这可是一个宏伟的人生愿景，也可是一个唾手可得的目标，幸福无疑是抽象的，从字面的理解应该是由于人的状态改进而产生的喜悦、满足、感恩和富足的心理感受，矩阵纵横自2010年初成立以来，无时无刻不在进行调整和改进自己的状态，这种改变让我们时刻充满着幸福的感受，无论对于设计还是生活，我们将一如既往地自我改变和超越。在当今中国高速发展的大环境下，设计在当今虽然还有许多不如意之处，但是也在逐渐向良性的方向发展，矩阵团队也得益于发展的大浪潮。我们大踏步向前，却不至于乱了阵脚，我们勤奋耕耘，却有意放慢脚步回顾和整理思绪。设计是商业化的艺术，我们乐此不疲地在商业与艺术之间寻找平衡点。方向明确在当下社会显得弥足珍贵，矩阵纵横从容走过人生的前五年，迈向充满未知却有无限想象的下一个五年，朝着我们的梦想前行。"回归东方"是我们的终极目标，这种回归是个体、团队、社会的呼喊，大势所趋。自从百余年前中国大门被撬开之后，中国一直被西方文明所"绑架"，在不断学习，不断地迷失和抛弃，我们舍近取远，却换回来的是不自信、不确定。这就是我们的现状，也是必然的结果，因为那不是自然的生长，浮于表象，不是我们血液里面流淌的东西。让生活回归；让审美价值回归；让设计回归，回归自然。就像欧美人一直推崇的东方智慧"侘寂"（WABI-SABI），应该是朴素、寂静、谦逊、自然，生活如斯，设计更是如此，只有心静方能从容。或许我们的目标还需要更多时间去证明，但是它很明确，很坚定。五年中我们完成了百余个国内精品项目，矩阵的今天应该感谢信任支持我们的新老客户！感谢一路走来不离不弃的团队小伙伴！我个人也是从这些年与团队的一步步成长，从稚嫩走向成熟，也得益于二位亦师亦友的搭档王冠和王兆宝先生，我们给自己一个承诺，坚信梦想能够实现，借用马云一句名言：梦想还是要有的，万一实现了呢:)

2015.01.12 刘建辉

Two or three things to remember

One night in early March 2010, I got a call "Do you have wine?" I said, "Yes. Come and get it." After taking the wine, he went to the door but turned around, "I am setting up a company with Idmen. You wanna join?" I agreed. He said, "I am talking to Idmen and will get back to you when confirmed." Soon I got the call, saying "Let's do it on 16th this month."

Once we three got drunk after a party and hung out. He said, "So glad I have you two." I said, "Don't mention it. Even 10 years is just the beginning of our business."

I proposed System of Partnership at the shareholders meeting and it was smoothly passed. After that Matrix established the System of Partnership.

Five years went by and another five years is forthcoming. At the summing up meeting, I said, "You two may say something about our future." Happy and calm, yes, that's it. Hope our partners will obtain self-fulfillment and happy family. Design calmly and follow our heart.

2015.1.17　Zhaobao Wang

二三小事记

2010年3月初的一个晚上，我接到电话："你那还有酒吗？我来拿。"我说："有，来吧。"拿了酒，他走到门口却回头问："哎，我和建辉攒一公司，就上次一起喝酒那哥们，要一起吗？"我说好呀。他说："那我先问下建辉，这公司是建辉和我攒的，得先问问他。"我说："好呀，要定了你告诉我就行。"一会接到电话说："准备下，那咱就这月16号开始吧。"

有次聚会酒喝多了哥仨一起压马路，他说："今天很高兴，得亏你哥俩。"我说："这才哪到哪，我们一起的事业是10年起步的。"

股东会我提出讨论下合伙制的事。这有什么好讨论的，赶紧制定规则执行呀。要是以前有人和我们谈这事我们还不乐疯了。于是矩阵确立合伙人制。

春去秋来，转眼一个五年逝去，又迎接一个新的五年，总结会上，我说："你哥俩梳理下我们的愿景吧。"幸福从容，对，幸福、从容。希望小伙伴们实现自我、家庭幸福，从容地进行设计，不做有违自己内心的事。

2015.1.17　王兆宝

CONTENTS 目录

001	Vanke Yue Bay Sales Center		001	万科悦湾销售中心
015	Fu Chun Oriental Sales Center		015	富春东方销售中心
025	Zhongtai Tiancheng Office		025	中泰天成办公室
037	Vanke Yue Bay Duplex Western Sample House A2		037	重庆万科悦湾A2复式洋房
051	Xiangdi Garden B-type Show Flat, Galaxy, Tianjin		051	星河天津香堤园B户型样板房
061	Chengdu Forte 606 — 607 Type Office Show Flat		061	成都复地606-607户型办公样板房
071	Chengdu Forte 608 Type Office Show Flat		071	成都复地608户型办公样板房
079	Chengdu Forte 609 Type Office Show Flat		079	成都复地609户型办公样板房
087	Shenzhen St. Morris Building C11 Show Flat Unit 22A		087	深圳圣莫丽斯C11栋一单元22A样板房
097	Chongqing Vanke Yue Bay 401 Type Show Flat		097	重庆万科悦湾401户型样板房
109	Chongqing Vanke Xijiu Sales Center		109	重庆万科西九销售中心
121	Zhengzhou Vanke Town Sales Center		121	郑州万科城销售中心
133	Chongqing Vanke Town DS Type Villa		133	重庆万科城DS户型别墅
151	Chongqing Vanke Town LP6 Type Villa		151	重庆万科城LP6户型别墅
167	Chengdu Vanke Office Building		167	万科地产成都办公室
187	Chengdu Vanke Jinyu Tixiang Sales Center		187	成都万科金域缇香销售中心
201	Guiyang Junfa Office Sample House		201	贵阳俊发办公样板房
211	Quanzhou Wandao Ziyuntai Sales Club		211	泉州万道紫云台销售会所
229	Quanzhou Wandao Office		229	泉州万道办公室
241	Chongqing Vanke Town Villa		241	重庆万科城别墅
251	Road King Group X1 Type Sample House, Changzhou		251	路劲集团（常州）X1户型样板房
263	Road King Group X2 Type Sample House, Changzhou		263	路劲集团（常州）X2户型样板房
275	Road King Group Y1 Type Sample House, Changzhou		275	路劲集团（常州）Y1户型样板房
291	Chongqing Vanke Yue Bay Villa		291	重庆万科悦湾别墅
305	Shenzhen Pengcheng Dadongcheng Villa		305	深圳彭成大东城别墅
319	Shenzhen GYENNO Technology Office		319	深圳臻络科技办公室
329	Ji'nan Road King Sales Center		329	济南路劲销售中心
337	Road King Group X6 Type Sample House, Changzhou		337	路劲集团（常州）X6户型样板房
347	Ji'nan Road King F-Type Sample House		347	济南路劲F户型样板房
361	Hefei Vanke Forest Park Building 17 One-Floor Show Flat		361	合肥万科森林公园17栋一层样板房
375	Hefei Vanke Forest Park Building 17 Two-Floor Show Flat		375	合肥万科森林公园17栋二层样板房

万科悦湾销售中心

重庆 / 中国

该项目将亚洲元素植入现代建筑语系，将传统意境和现代风格对称运用，用现代设计来隐喻中国的传统。水曲柳屏风与深色麻石的搭配既传统又流行，而且为空间营造出了充满魅力的对称感，使整个空间更具立体感，在美观之余，更增韵味，彰显东方的古典优雅气质。

ANDREW MARTIN
万科悦湾销售中心荣获2013—2014年英国安德鲁马丁大奖

APIDA
万科悦湾销售中心荣获2013年第21届香港亚太室内设计大奖——卓越奖

VANKE YUE BAY SALES CENTER

CHONGQING / CHINA

Planting Asian elements into modern architectural system, balancing traditional artistic conception with modern style, this project successfully integrated the new and the old, making the modern design a great metaphor of Chinese traditions. Chinese White Ash screen with Dark Granite is a traditional but popular match which creates a glamorous spacial symmetry, showing a more stereoscopical space not only beautiful but also charming with the unique oriental classic elegance.

ANDREW MARTIN
Vanke Yue Bay Sales Center got Excellence of 2013—2014 British Andrew Martin award

APIDA
Vanke Yue Bay Sales Center got Excellence of 2013 Hong Kong APIDA award

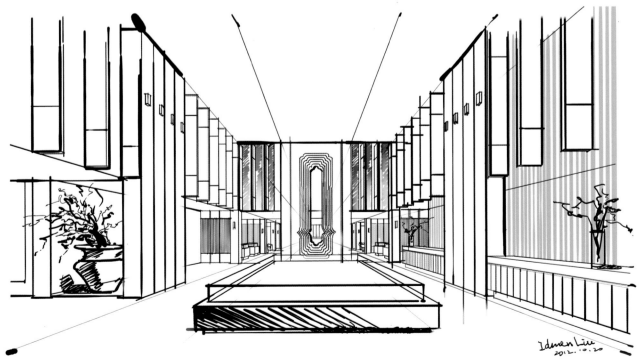

011

01	Waterscape	
02	Reception Desk	
03	Reception Hall	
04	Open Office	
05	Meeting Room	
06	Gallery	
01	水景	
02	接待前台	
03	接待前厅	
04	开敞办公室	
05	会议室	
06	艺术连廊	

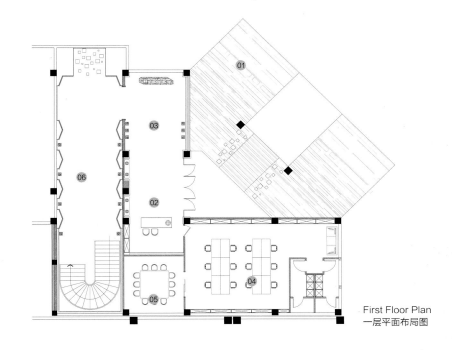

First Floor Plan
一层平面布局图

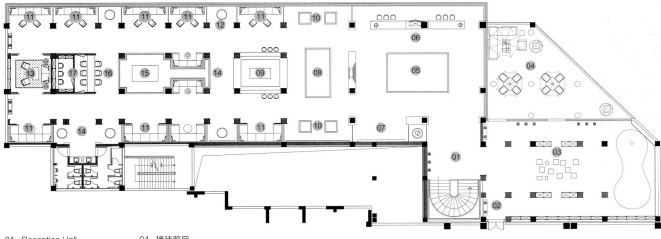

01	Reception Hall	01	接待前厅
02	Open Sample House	02	通向样板房
03	Golf Theme Pavilion	03	高尔夫主题馆
04	Outdoor Terrace	04	户外露台
05	Main Sand Table	05	主沙盘
06	Sales Control Service Desk	06	销控服务台
07	Vertical Sand Table	07	立面沙盘
08	Model Display Area	08	模型区
09	Bar	09	水吧区
10	Condo Model	10	分户模型
11	Discussion Area	11	洽谈区
12	Hallway Landscape	12	过厅景区
13	VIP Room	13	VIP室
14	Hallway	14	过厅
15	Water Scene Lounge	15	水景休闲区
16	Contract Area	16	签约区
17	Financial Contract Room	17	财务合同室

Second Floor Plan
二层平面布局图

013

富春东方销售中心

安庆 / 中国

在设计上考虑建筑与室内空间的整体性，用干净简单的块面来处理空间，古木纹石材与黑钢板的雕花的应用，使得空间氛围朴实，既有文化又有现代感。软装配饰上，经过改良的台灯及烛台、精致的陶罐，将空间以叙述的方式串联起来。

APIDA
富春东方销售中心荣获2013年第21届香港亚太室内设计大奖

FU CHUN ORIENTAL SALES CENTER

ANQING / CHINA

Considering the integrality of construction and interior space, we chose plain and simple block surface in space arrangement. The application of Black Forest Marble and Black Steel Carving gives the space a simple, artistic yet modern atmosphere. Decorative accessories like modified desk lamps, candlesticks and delicate pottery pots connect every dispersive corner into an entirety.

APIDA
Fu Chun Oriental Sales Center got Excellence of 2013 Hong Kong APIDA award

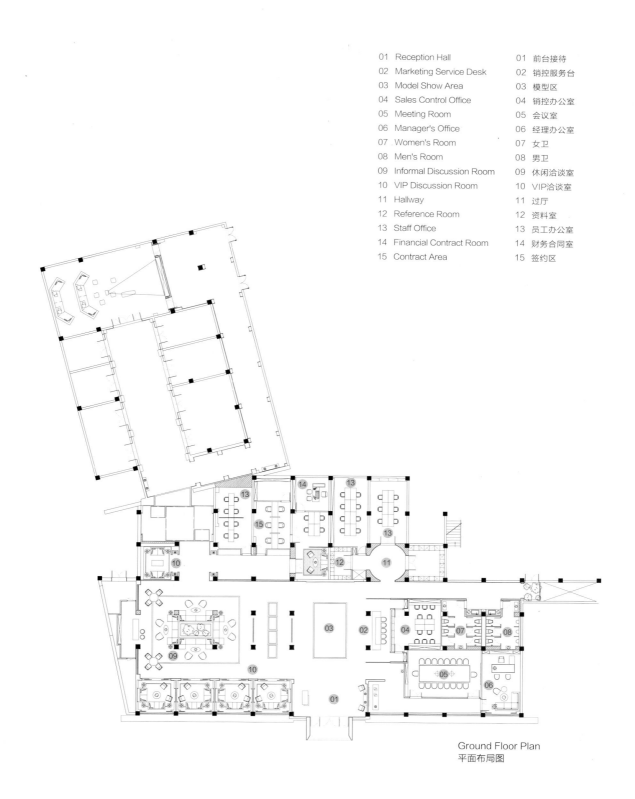

01	Reception Hall	01	前台接待
02	Marketing Service Desk	02	销控服务台
03	Model Show Area	03	模型区
04	Sales Control Office	04	销控办公室
05	Meeting Room	05	会议室
06	Manager's Office	06	经理办公室
07	Women's Room	07	女卫
08	Men's Room	08	男卫
09	Informal Discussion Room	09	休闲洽谈室
10	VIP Discussion Room	10	VIP洽谈室
11	Hallway	11	过厅
12	Reference Room	12	资料室
13	Staff Office	13	员工办公室
14	Financial Contract Room	14	财务合同室
15	Contract Area	15	签约区

Ground Floor Plan
平面布局图

中泰天成办公室

深圳 / 中国

项目位于深圳市南山区中泰天成办公楼第16层，其主要功能是售楼与办公样板间展示于一体的空间。它是一位稳重的、讲究细节的都市精英，在橡木色的底子下又跳跃着一个年轻而充满活力的绿色小清新。整个空间主要分为销售区和办公样板区两大块，两者间没有明显的隔墙划分，而是通过各功能区造型的延续，用通透的金黄色玻璃隔断，让各区域相互独立又互相贯通，且五米多的层高通过局部夹层的处理让空间层次更为突出，使用功能更为多样。

APIDA

中泰天成办公室荣获2013年第21届香港亚太室内设计大奖

ZHONGTAI TIANCHENG OFFICE

SHENZHEN / CHINA

The project is located in 16 F of Zhongtai Tiancheng Office Building, Nanshan District, Shenzhen. It's a functional space integrating sales center and model office. It is a mature urban elite particular about details, yet under the appearance of oak color is a young and lively freshness beauty shining and dancing. The whole space is divided into two parts as sales zone and model office zone; instead of wall partitions, functional zones are separated by their natural structure and crystal golden glass which makes every part independent and connected. A few inter-layers between the 5 meter space make more sense of layering and variety of functions.

APIDA

Zhongtai Tiancheng Office got Excellence of 2013 Hong Kong APIDA award

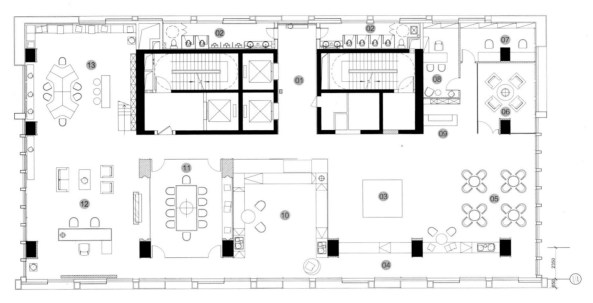

Ground Floor Plan
平面布局图

01 Elevator Hall   01 电梯厅
02 Washroom   02 卫生间
03 Model Show Area   03 模型展示区
04 Service Area   04 服务区
05 Discussion Area   05 洽谈区
06 Discussion Room   06 洽谈室
07 Financial Room   07 财务室
08 Manager Office   08 经理室
09 Water Bar   09 水吧
10 Transparent Garden   10 透明花园
11 Meeting Room   11 会议室
12 Chairman Office   12 董事长办公室
13 Workspace   13 办公区域

重庆万科悦湾A2复式洋房

重庆 / 中国

"新亚洲"的风格定位,"复古""奢华"在这个作品中得到很好的体现,在大面积灰色调的烘托下,深色木饰面与细节处金漆的勾勒;实木断面与绒布硬包的材质对比,突出了空间与众不同的气质。陈设软装在整个空间中起到了画龙点睛的作用。

APIDA

重庆万科悦湾A2复式洋房荣获2013年第21届香港亚太室内设计大奖

VANKE YUE BAY DUPLEX WESTERN SAMPLE HOUSE A2

CHONGQING / CHINA

Style orientation— "New Asia" . "Retro" & "Luxury" are fully presented in this work. With an atmosphere of gray, the delicate dark wood veneer and subtle golden paint, as well as a contrast between solid wood section and lint metope hard roll, the entire space reflects a unique taste. And the displaying soft decorations to the space are the very finishing points.

APIDA

Vanke Yue Bay Duplex Western Sample House A2 got Excellence of 2013 Hong Kong APIDA award

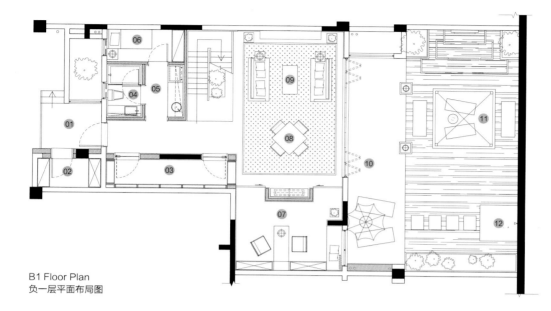

B1 Floor Plan
负一层平面布局图

01 Garage Entrance
02 Storage
03 Red Wine Cellar
04 Public Washroom
05 Workroom
06 Servant Room
07 Study
08 Chess Room
09 Family Room
10 Sightseeing Balcony
11 Yoga & Meditation Room
12 Shower Area

01 车库入户
02 储物间
03 红酒窖
04 公卫
05 工作间
06 保姆房
07 书房
08 棋牌室
09 家庭厅
10 景观阳台
11 瑜伽冥想区
12 淋浴区

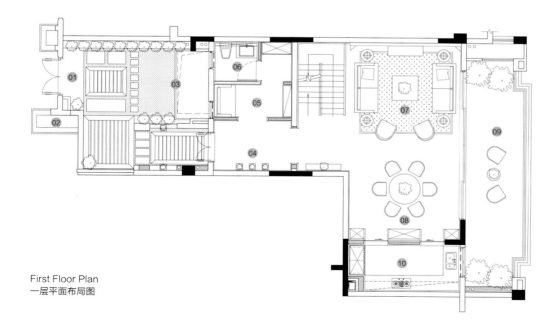

First Floor Plan
一层平面布局图

01 Front Yard
02 Flower Bed
03 Sunk Courtyard Upper Space
04 Hall
05 Shoes Changing Area
06 Public Washroom
07 Living Room
08 Dining Room
09 Sightseeing Balcony
10 Kitchen

01 入户庭院
02 花池
03 下沉庭院上空
04 玄关
05 换鞋区
06 公卫
07 客厅
08 餐厅
09 景观阳台
10 厨房

星河天津香堤园B户型样板房

天津 / 中国

XIANGDI GARDEN B-TYPE SHOW FLAT, GALAXY, TIANJIN

TIANJIN / CHINA

这是一套展示性的样板房空间，田园风格一般给人的印象是粗狂、低廉的感觉，设计师在这个案例上，尽量去改变一般人的思维定式，在硬装的油漆及地面的材料的选择上下了些功夫，没有选用以往开放的油漆面，而是改成精致的封闭漆面。软装饰选用了花鸟的主题布艺，摈弃了原来常见的格子布艺，家具的漆面则使用了最近流行的水洗白漆面，把时尚的元素融入质朴的空间，达到出奇的视觉效果。

This is a set of model house for exhibiting. Common impression of Country Style is cheap and rough; the designer made lots of efforts in the selection of painting and floor materials, trying to change people's fixed impression of Country Style. Instead of the old open paint finish, more delicate closed paint finish is used here. Fabrics with flowers and birds rather than plaid fabric are chosen for soft decorations, while the popular White Washed paint finish is used in furniture, integrating fashion elements into simple space for a magical visual effect.

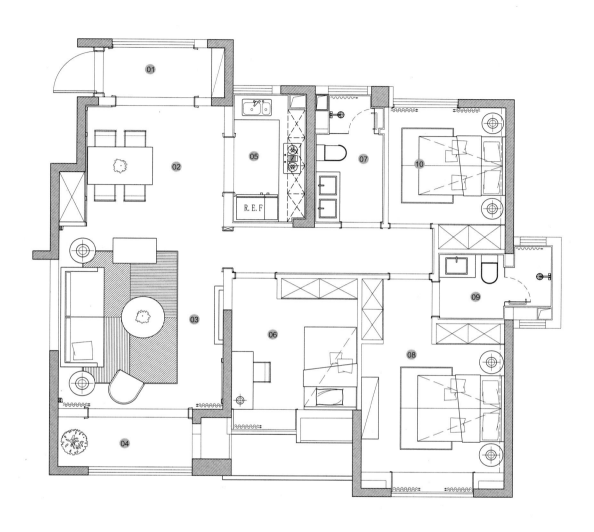

01	Entrance Hall	01	玄关
02	Dining Room	02	餐厅
03	Living Room	03	客厅
04	Balcony	04	阳台
05	Kitchen	05	厨房
06	Children's Room	06	儿童房
07	Washroom	07	洗手间
08	Master Bedroom	08	主卧室
09	Main Bathroom	09	主卫
10	Guest Bedroom	10	客卧

Ground Floor Plan
平面布局图

成都复地606-607户型
办公样板房

成都 / 中国

这是一个创意型办公样板房，也是有别于传统的办公空间，首先在平面布局上做了些创新，把所有功能全部放置在中间位置区域，如：前台接待、迷你吧、员工休息位，还有打印复印区，这样在使用起来非常方便高效；然后在色调上完全使用了黑白灰的一个基调搭配着绿色植物及暖色灯光，使得空间看起来宽敞明亮，对空间的展示起到了加分的作用，更好地有利于销售。

CHENGDU FORTE 606—607 TYPE OFFICE SHOW FLAT

CHENGDU / CHINA

This is a creative model office room. Different from traditional office space, all the functional facilities are placed in the middle, such as reception desk, mini bar, staff rest area and printing-copy area. The main color style is black, gray and white with some green plants and warm lights, which makes the space larger and brighter.

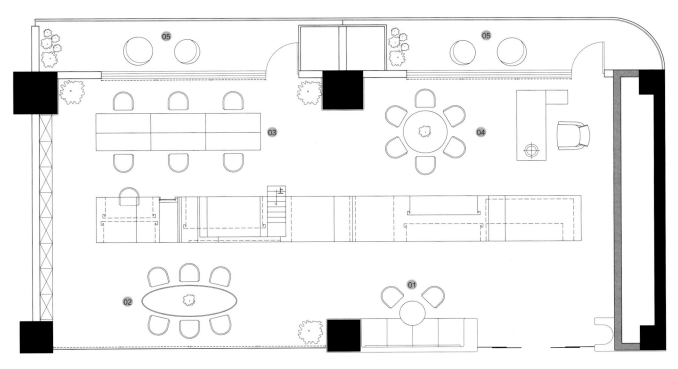

Ground Floor Plan
平面布局图

01 Reception Area   01 接待区
02 Meeting Room   02 会议室
03 Open Workspace   03 公共办公区
04 Head Office   04 主管室
05 Outdoor Lounge   05 户外休闲区

CHENGDU FORTE 608 TYPE OFFICE SHOW FLAT

CHENGDU / CHINA

成都复地608户型办公样板房

成都 / 中国

成都是个休闲趣味十分浓厚的城市，人们悠闲地享受生活，不像北上广深的繁忙而快节奏的生活。所以在拿到这个案子的时候要深入剖析当地的特色，了解需求，尽量在50平方米的空间里诠释成都人对工作空间的向往。

新亚洲风格主张以具有浓厚地域特色的传统文化为根基，把亚洲元素植入现代建筑语系，将传统意境和现代风格对称运用。在关注现代生活舒适性的同时，让当地传统文化得以传承和发扬。在本案例中，设计师在布局上运用对称的空间寻求一种曲径相通的效果，在色调上传承了当地建筑的特色，采用大面积留白的手法，给人带来素雅安逸的状态。造型上简洁刚毅又不失细节和品位。其风格家具以实木为主，并注重实用性。

清晨阳光通过全落地大窗照射在办公区带有传统纹样的条桌上，其主人享受其中。休息区舒适的软垫、风情的而禅意的青竹饰品，让本身呆板的空间添加了不少意境。素雅的洽谈区环境配上舒适的座椅尽显温馨惬意。

Chengdu is a city of leisure and fun where people enjoy life leisurely, different from the busy life in Beijing, shanghai, Guangzhou and Shenzhen. So we need to do a deep analysis on local culture, learn their demands on this case, and try the best to fulfill Chengdu people's expectations toward office space within the 50 m^2 area.

New Asia style advocates injecting Asian factors into modern architectures and conducting symmetric use of traditional artistic conception and modern style on the basis of presentation of local culture and regional characteristics. Thereby, local traditional culture is spreading while achieving the comfort of modern life. In this case, the designer uses symmetric structure to realize connection in meandering space, and for colors, he keeps the feature of local architectures, using large white space to bring a feeling of simplicity, elegance and leisure. The structure is solid with taste of details, and wood furniture is of practical functions.

Cozy cushion and the Buddhist bamboo decorations in rest area, simple and elegant negotiate area with comfortable chairs, all these surroundings create a perfect environment. What a enjoyable moment it will be when morning sunshine goes through French windows and lies on the office table with traditional fixtures.

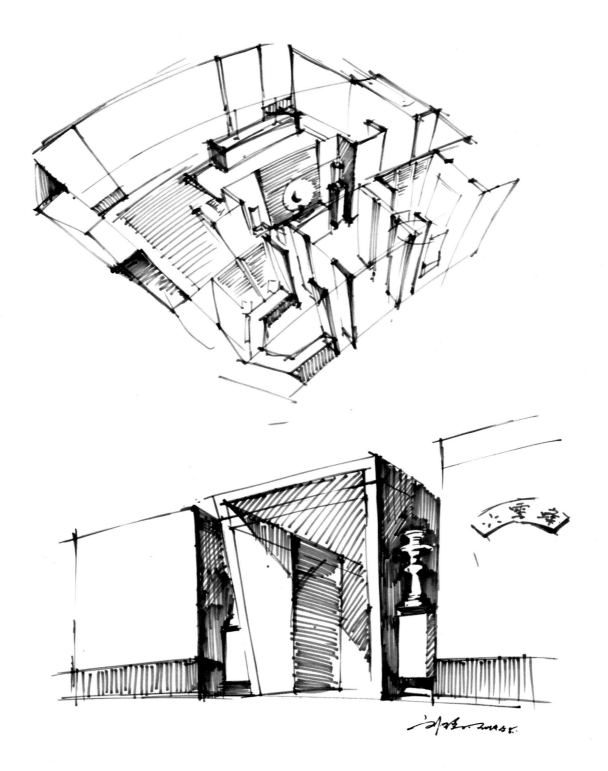

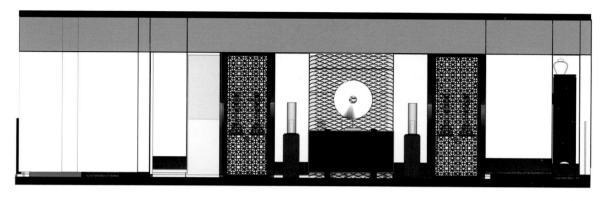

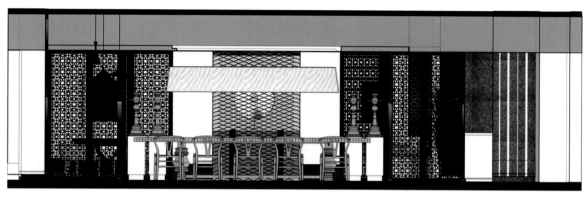

01 Reception Area
02 Discussion Area
03 Workspace

01 接待区
02 洽谈区
03 办公区

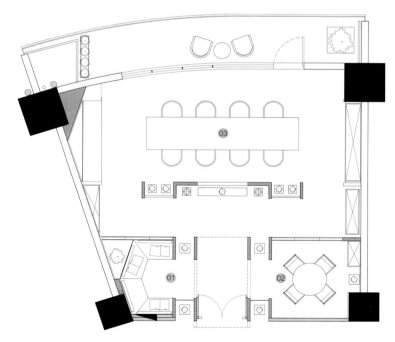

Ground Floor Plan
平面布局图

CHENGDU FORTE 609 TYPE OFFICE SHOW FLAT

CHENGDU / CHINA

成都复地609户型办公样板房

成都 / 中国

此次办公样板房的项目设计旨在提供一种实用而不乏创意的办公空间解决方案。平面布局上将休闲区、会客区、经理办公区以公共办公区为中心发散,减少人流动线交叉,提高工作效率。入口以斑斓的挂画拉开空间,象征探索与激情的山地车作为装饰元素置于入口。内部空间主色调以白、绿为主,墙上的树状书架成为空间的一大亮点,使整个空间清新而富有活力。极具动感的线条由天花造型延伸至墙面,丰富了空间的层次,使空间更为灵动。空间的点睛之笔在于会客区,狭长的空间配以黄绿色调的变化拼接,中间以白色自行车作为桌脚,人行其中如同林间漫步,打破了办公空间原本沉闷的气氛。

本方案在解决小型空间内的多功能实现的同时,将强烈的设计感官引入其中。功能与审美相结合,带给客户群体更为独特的体验。

This project is targeted to provide an office space solution of both practical usage and creativity. Leisure area, customer area and manager office area are set around the central public office area, which disperses office staff moving lines and avoids mutual interference, thereby improving working efficiency. A colorful painting on the entrance wall opens the whole space, and a mountain bike symbolizing exploring and passion is decorated near the entrance. Main colors of interior space are white and green; the tree-shaped book shelf on the wall becomes a very highlight of the space and makes it fresh and energetic. Dynamic lines extended from ceiling to walls enrich levels of space. The most special part that can be called as the finishing touch lies in the customer area, a long space colored in gradient yellow-green, where white bikes are made as table bases in the middle part, and people walking there feel like walking in the forest and enjoy the free atmosphere.

This case realizes multi-functions in small space while injecting a strong sense of designing. Function and aesthetic are perfectly combined and bring customers very unique living experience.

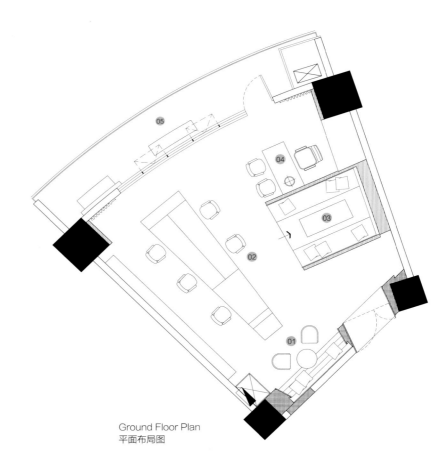

01　Entertainment Area
02　Open Workspace
03　Meeting Area
04　Manager Office
05　Outdoor Terrace

01　休闲区
02　开敞办公区
03　会议区
04　经理办公室
05　阳光休闲区

Ground Floor Plan
平面布局图

深圳圣莫丽斯C11栋一单元22A样板房

深圳 / 中国

SHENZHEN ST. MORRIS BUILDING C11 SHOW FLAT UNIT 22A

SHENZHEN / CHINA

精练的布局、黄金的分割。在完整的功能与使用的基础上,设计者运用新中式的现代手法,让整个空间格外明朗而富有情趣。黑白灰的主色调下,细节和纹理极为考究,雅士白、亚光黑、氟碳漆……或者是简化的中式雕花组合、回纹的重复阵列、橡木平和的肌理和雅致的色调……都彰显了平稳大气的新中式典范。在开阔干净的基调上,软装在柔化空间的同时,更注重品质和质感的表达。这是完美化生活的展现,也是对精英生活的尽情升华。

Delicate and concise layout, sections in golden ratio. Designers use modern new Chinese style to make the entire space rather bright, broad and full of taste. With main colors as black, white, grey, details and textures are very exquisite, using Ariston white, Matte black, Fluorocarbon paint... or the simplified Chinese wood carvings, the repeat array of frets, soft oak texture and elegant hues... all demonstrating the stable and profound paradigm of new Chinese style. On the basis of a broad and clear space, soft decorations soften the space and meanwhile manifest quality and sense. This is a presentation for perfect life and a sublimation for elite life.

01	Entrance Hall	01	入户玄关
02	Outdoor Landscape	02	户外景观
03	Living Room	03	客厅
04	Western Bar	04	西餐酒吧区
05	Dining Room	05	餐厅
06	Kitchen	06	厨房
07	Living Terrace	07	生活阳台
08	Second Bedroom	08	次卧
09	Washroom	09	卫生间
10	Hallway	10	过厅
11	Children's Room	11	儿童房
12	Family Bathroom	12	公卫
13	Study	13	书房
14	Cloakroom	14	衣帽间
15	Master Bedroom	15	主卧
16	Master Bathroom	16	主卫
17	Terrace	17	露台
18	Balcony	18	阳台

095

CHONGQING VANKE YUE BAY 401 TYPE SHOW FLAT

CHONGQING / CHINA

重庆万科悦湾401户型样板房

重庆 / 中国

山城重庆是一个有着深厚文化底蕴的城市，房屋大多依山而建。山城雨多雾大，常年生活在这样的环境里，人们渴望明亮宽敞的居住空间。新中式的淡雅明快的色调，很快地捕捉到客户的需求。其风格造型简洁，多以修边套线强调空间感。色调柔和明快。由于表现风格简洁大气，所以对细节收口要求更高，更能体现产品的品位和档次。

客、餐厅空间连通为一体，彰显其户型特色，家具摆放紧凑使空间饱满舒适。主人房运用中式花格手法装饰床头背景，镂空花格实虚结合体现空间层次感。巧用窗台让空间得以拓展延伸。书房、客房也沿用同样的色调搭配，与其他空间相得益彰……

Mountain city Chongqing is a city with profound culture and history. Houses here are mostly built near mountains. It rains a lot and usually has heavy fog here. People in this environment desire bright and broad living space. Elegant and sprightly tone of New Chinese style just satisfies customers' requirements. New Chinese style is with simple structure and usually emphasizes space sense with trimming and pallial lines. And it has very high standard on details and therefore more obvious to see the quality and taste.

Living room is connected with dining room, and furniture is well-organized in limited space. Bedside background in the ower's bedroom is decorated with Chinese lattices; hollow lattice promotes a sense of layering. Windows are skillfully used to extend entire space. Study room and guest room are decorated with the same color tone, just in perfect match with all the other rooms.

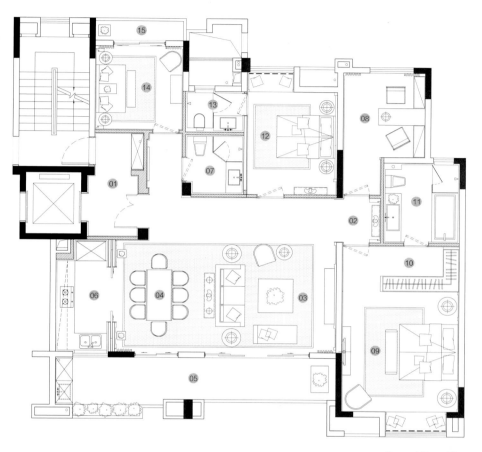

Ground Floor Plan
平面布局图

01	Elevator Hall	01	电梯厅
02	Hallway	02	玄关
03	Living Room	03	客厅
04	Dining Room	04	餐厅
05	Sightseeing Balcony	05	景观阳台
06	Kitchen	06	厨房
07	Family Bathroom	07	公卫
08	Study	08	书房
09	Master Bedroom	09	主卧室
10	Cloakroom	10	衣帽间
11	Main Bathroom	11	主卫
12	Children's Room	12	亲子房
13	Second Bedroom	13	次卫
14	Multifunction Room	14	多功能房
15	Living Balcony	15	生活阳台

重庆万科西九销售中心

重庆 / 中国

CHONGQING VANKE XIJIU SALES CENTER

CHONGQING / CHINA

一座集装箱搭建的建筑，平躺在重庆繁华的都市中，看似冰冷、生硬、无情，和周围的一切都格格不入。但我们给它披上了一件草绿色的外衣，给予它一种新的个性和生命力。它的内部和外在一样随处可见绿色。我们并没有改变它原始粗犷的性格。墙面天花的波纹钢板，都让它尽情地暴露，尽情地发泄……

A building made of containers, lies quietly in the prosperous city center of Chongqing. Cold, stiff, heartless and incompatible with all around as though it seems, but when dressed with a lawn green outwear, there it stands with its unique feature and vigor. Green is everywhere inside the building as it does in the outside; however, we did not change its originality and roughness. Corrugated steel of the container is kept what it was as walls and ceilings, exposing naturally and wildly.

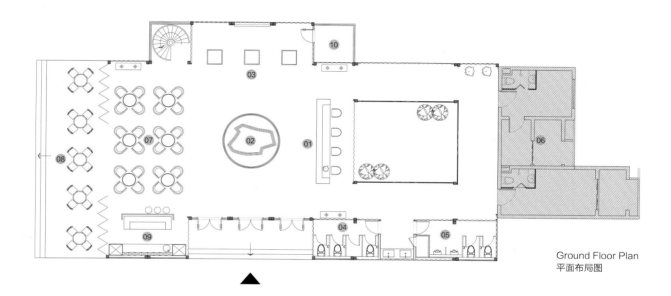

01	Service Desk	01	服务台
02	Large Sand Table	02	大沙盘
03	Condo Model Drawing	03	分户模型图
04	Women's Room	04	女洗手间
05	Men's Room	05	男洗手间
06	Sample House Area	06	样板房区域
07	Discussion Area	07	洽谈区
08	Lounge Area	08	休息区
09	Water Bar	09	水吧区
10	Warehouse	10	库房

Ground Floor Plan
平面布局图

郑州万科城销售中心

郑州 / 中国

大空间的大作为

设计师对整体室内结构进行解构和重组，满足诸多功能以外，力求整体的和谐与开阔。穿过低调内敛的前厅，恢宏大气的中庭让人眼前一亮、豁然开朗。核心筒钻石亮面的外衣，构建整个会所的建筑之美，时尚、大气，不乏庄重与质感，给予核心筒新的生命力和号召力。

ANDREW MARTIN

郑州万科城销售中心荣获2013—2014年英国安德鲁马丁大奖

ZHENGZHOU VANKE TOWN SALES CENTER

ZHENGZHOU / CHINA

Great accomplishment in large space

Designers reconstructed and regrouped the interior structure to create a harmonious and wide open space and meanwhile achieve more functions. Across the low-key antechamber, a magnificent atrium suddenly came into view and presented a broad clear space. Shining diamond outwear of the core tube gave it new life and power, representing the architecture beauty of the club, fashionable, generous and majestic.

ANDREW MARTIN

Zhengzhou Vanke Town Sales Center got Excellence of 2013—2014 British Andrew Martin award

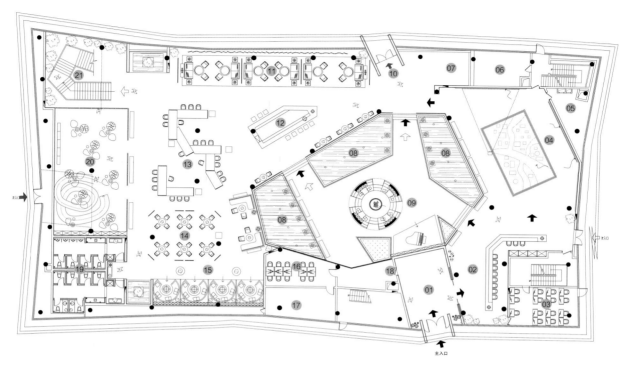

Groung Floor Plan
平面布局图

01	Entrance Hall	01	入口前厅
02	Reception Desk	02	前台接待
03	Customer Service Call Center	03	客服CALL客室
04	Model Show Area	04	模型展示区
05	Ventilator Room	05	风机房
06	Storage	06	物业储藏室
07	Technique Show Area	07	工法展示区
08	Atrium Waterscape Area	08	中庭水景区
09	Atrium	09	中庭
10	Aisle to Sample House	10	去样板房区
11	Primary Customer Discussion Room	11	初步客户洽谈区
12	Marketing Control Area - Bar Service Desk	12	水吧服务台销控区
13	Customer Experience Area	13	客户体验区
14	Customer Discussion Area & Lounge	14	客户洽谈区及休息区
15	Depth Customer Discussion Area	15	深度客户洽谈区域
16	Marketing Office	16	销售办公室
17	Fire Control Room	17	消防控制室
18	Material Storage	18	物料储藏室
19	Men and Women's Washroom	19	男女洗手间
20	Discussion Area (Children Allowed)	20	带儿童洽谈区
21	Staircase	21	楼梯间

CHONGQING VANKE TOWN DS TYPE VILLA

CHONGQING / CHINA

重庆万科城DS户型别墅

重庆 / 中国

This project applies both traditional conception and modern style, and briefly presents oriental symbols with modern design method. With no redundant modeling or carvings, the match of pure dark ash tree veneer and light linen wallpaper shared both traditionality and popularity, and created for the space a glamorous classical atmosphere which made it more stereoscopic and charming, showing a unique feeling of oriental style.

本项目将传统意境和现代风格对称运用，用现代设计手法将东方的语汇去繁从简表达出来。没有多余的造型和雕花，纯粹的深色水曲柳饰面、浅色麻面墙布的搭配既传统又流行，而且为空间营造出了充满魅力的古典美，使整个空间更具立体感，在美观之余，更增韵味，彰显东方气质。

APIDA

重庆万科城DS户型别墅荣获2014年第22届香港亚太室内设计大奖——金奖

APIDA

Chongqing Vanke Town DS Type Villa got Gold Medal of 2014 Hong Kong APIDA award

01 Informal Living Room
02 Water Bar
03 BBQ Recreation Activity Room
04 VOID Waterscape
05 Entrance Side Hall
06 Garage
07 Family Bathroom
08 Feature Display Area
09 Staff Room
10 Laundry

01 休闲会客区
02 水吧区
03 BBQ娱乐活动室
04 VOID水景
05 入户偏厅
06 车库
07 公卫
08 特色展示区
09 工人房
10 洗衣区

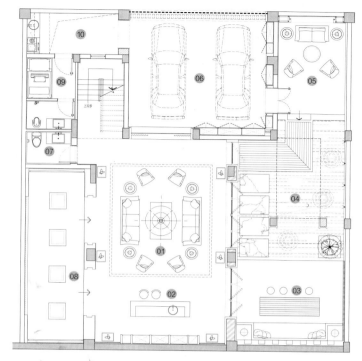

B1 Floor Plan
负一层平面布局图

01 Hallway
02 Living Room
03 Kitchen
04 Dining Room
05 Staircase
06 Washroom
07 Guestroom
08 Cloakroom

01 玄关
02 客厅
03 厨房
04 餐厅
05 楼梯间
06 卫生间
07 客卧
08 衣帽间

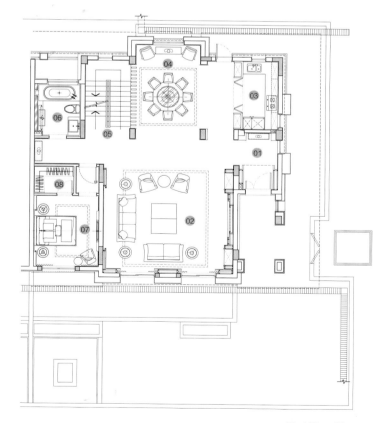

First Floor Plan
一层平面布局图

01 Hallway
02 Guestroom 1
03 Guestroom 2
04 Washroom
05 Cloakroom
06 Over the Living Room

01 多功能厅
02 客房一
03 客房二
04 卫生间
05 衣帽间
06 客厅上空

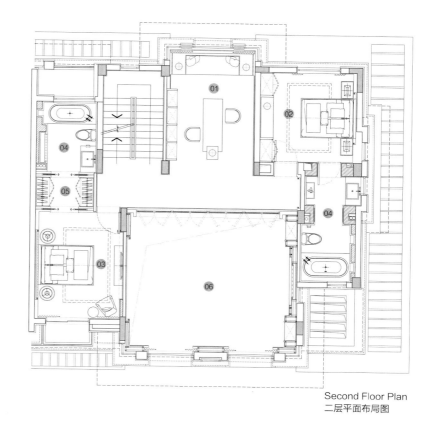

Second Floor Plan
二层平面布局图

01 Hallway
02 Master Bedroom
03 Cloakroom
04 Washroom
05 Study

01 内玄关
02 主卧室
03 衣帽间
04 卫生间
05 书房

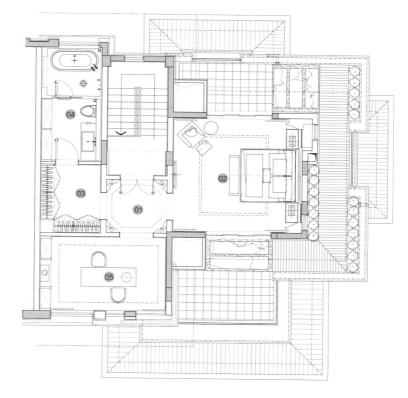

Third Floor Plan
三层平面布局图

重庆万科城LP6户型别墅

重庆 / 中国

低调奢华的空间，用现代简洁的手法表现。

整体的功能分区动静分明、流线畅然。负一层用作娱乐派对，为互动与开放的空间；一层用作休闲起居；二层用作学习、休憩。落落大方的格局组合，让生活更是从容。其演绎出精致而丰富的现代精英生活，更是值得向往。

设计线条干净而凝练的表达，有力却不失细节和质感。同色系、同纹理的组合，为冰冷的空间注入温暖的语言，散发现代都市新贵尊贵的气质。设计师在材质和色彩的搭配上，十足的考究。尼斯木立体的纹理搭配黑镜钢的力量，再配以鳄鱼纹理的绒布硬包，低调干练，传达精致的感情。

CHONGQING VANKE TOWN LP6 TYPE VILLA

CHONGQING / CHINA

This project is low-key luxurious space, presented with concise modern design method.

Functional zones are clearly classified into public and private space by fluent structure lines. The underground floor is for entertaining party, an open space for interaction; Floor 1 is for leisure and living; Floor 2 is for study and rest. Clear and convenient structure makes life much easier. The delicate and colorful modern elite life it creates is more desirable.

The design lines are clear and concise, powerful yet with details and quality sense. Combination of tone on tone and texture on texture creates a warm feeling for the cold space, showing the noble temperament of modern city nobility. Designers are very deliberate on materials and colors. Lacewood 3D texture matches with black mirror steel, along with crocodile texture flannelette veneer, displaying a delicate feeling.

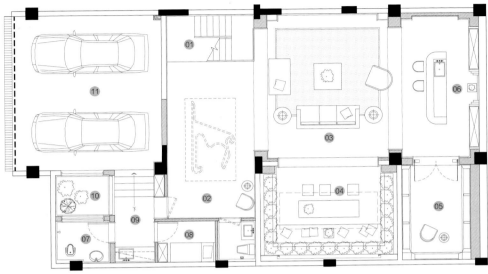

B1 Floor Plan
负一层平面布局图

01 Stairwell
02 Car Racing Theme Game Room
03 Multifunctrional Entertaining Area
04 Family-Gather BBQ Area
05 Red Wine Cellar
06 Bar
07 Washroom
08 Housemaid Room
09 Laundry Room
10 Interior Scenery
11 Garage

01 楼梯间
02 赛车主题游戏区
03 多功能娱乐厅
04 家庭聚会BBQ区
05 红酒窖
06 酒吧区
07 洗手间
08 保姆房
09 洗衣区
10 内景观
11 车库

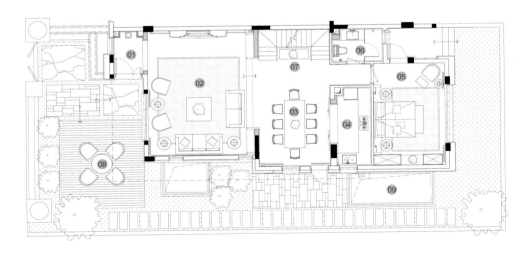

First Floor Plan
一层平面布局图

01 Entrance Hall
02 Living Room
03 Dining Room
04 Kitchen
05 The Elder's Room
06 Washroom
07 Staircase
08 Outdoor Scenery
09 Sunk Courtyard Upper Space

01 玄关
02 客厅
03 餐厅
04 厨房
05 老人房
06 洗手间
07 楼梯间
08 户外景观区
09 下沉庭院上空

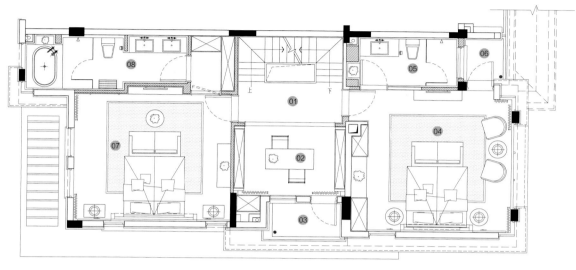

01	Stairwell	01	楼梯间
02	Study	02	书房
03	Terrace	03	露台
04	Second Bedroom	04	次卧室
05	Second Bedroom-Washroom	05	次卧室洗手间
06	Equipment Platform	06	设备平台
07	Second Master Bedroom	07	次主卧室
08	Second Master Bedroom-Washroom	08	次主卧室洗手间

Second Floor Plan
二层平面布局图

01	Stairwell	01	楼梯间
02	Cloakroom	02	衣帽间
03	Master Bedroom	03	主卧室
04	Master Bedroom-Washroom	04	主卧室洗手间
05	Terrace	05	露台

Third Floor Plan
三层平面布局图

165

CHENGDU VANKE OFFICE BUILDING

CHENGDU / CHINA

万科地产成都办公室

成都 / 中国

艺术性与功能性并存

设计理念为"传承、舒适、文化","以人为本"把人作为设计的重要参考物,重视环境人性化解决之道,强调空间合理布局及细节的关怀设计。在整个办公室设计中将一些本土地域文化融合其中,是此项目的最大亮点,也在以往刻板的办公空间设计中增加了些许创新精神。

Coexistence of artistry and functionality

"Inheritance, Comfort, Culture" is the design conception. "People First" sets people as an important reference in design, which attaches importance to humanized solution of environment and emphasizes reasonable space arrangement and details caring design. The brightest spotlight of this project is the blend of local regional culture into office design, which presented some creative spirit to the stiff office design before.

179

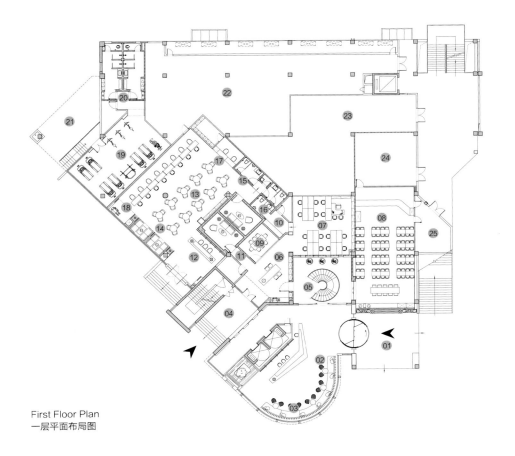

First Floor Plan
一层平面布局图

01	Main Entrance	01	主入口
02	Reception Area	02	前台接待区
03	Reception Lounge	03	接待休息区
04	Elevator Hall	04	电梯厅
05	Panoramic Lift	05	景观楼梯间
06	Entrance Hall	06	前厅
07	Customer Service Center	07	客户关系中心
08	Staff Cafeteria	08	员工餐厅
09	Meeting Room	09	会议室
10	Water Bar	10	水吧
11	Culture Wall	11	文化墙
12	Customer Reception Area	12	客户接待区
13	Contract Area	13	签约区
14	Sofa Waiting Area	14	沙发等候区
15	Cashier Desk	15	收银
16	Men & Women's Room	16	男女卫生间
17	Storage	17	储藏间
18	Printing Area	18	复印区
19	Fitting Room	19	健身房
20	Men & Women's Dressing Room	20	男女更衣室
21	Company Garage	21	公司库房
22	Company Archive	22	公司档案室
23	Material Display Area	23	材料展示间
24	Kitchen	24	厨房
25	Dishwashing Room	25	洗碗间

01	Entrance Hall on the 2nd Floor	01	二楼门厅
02	Sightseeing Elevator	02	景观楼梯间
03	Company Culture Wall	03	公司文化墙
04	Water Bar	04	水吧
05	Lounge	05	休闲区
06	Director Room 1	06	总监室一
07	Director Room 2	07	总监室二
08	Director Room 3	08	总监室三
09	Director Room 4	09	总监室四
10	Open Discussion Room	10	开敞讨论室
11	Discussion Room	11	讨论室
12	Open Office 1	12	开敞办公区一
13	Men's and Women's Room	13	男女卫生间
14	Open Office 2	14	开敞办公区二
15	Printing Room	15	文印室
16	Storage	16	库房
17	Open Office 3	17	开敞办公区三
18	Smoking Area	18	吸烟区
19	Small Meeting Room	19	小会议室
20	Medium Meeting Room	20	中会议室
21	Drawing Display Area	21	看图区
22	Casual Discussion Area	22	休闲讨论区
23	Material Display Area	23	材料展示区
24	Open Office	24	开敞办公区

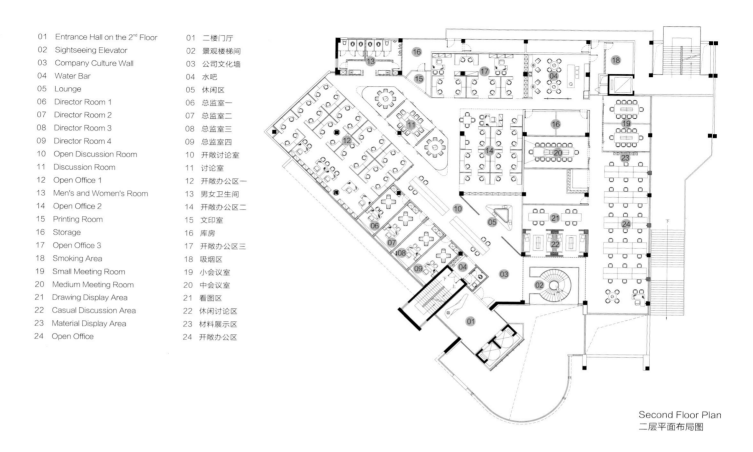

Second Floor Plan
二层平面布局图

01	Entrance Hall on the 3rd Floor	01	三楼门厅
02	Sightseeing Elevator	02	景观楼梯间
03	Company Culture Wall	03	公司文化墙
04	Water Bar	04	水吧
05	Washroom	05	洗手间
06	Open Negotiation Area	06	开放洽谈区
07	Alternate Office	07	备用办公室
08	Office 1	08	办公室一
09	Office 2	09	办公室二
10	General Manager Office	10	总经理办公室
11	Secretary Office	11	总经理秘书办公室
12	Storage	12	储藏室
13	Open Office	13	开敞办公区
14	Financial Management Department	14	财务管理部
15	Project Development Department	15	项目发展部
16	Director Office	16	总监
17	Alternate Office Booth	17	备用办公位
18	Project Purser Department	18	项目事务部
19	Financial Archives	19	财务档案室
20	Water Bar	20	水吧
21	Smoking Area	21	吸烟室
22	Computer Room	22	电脑机房
23	Small Meeting Room	23	小会议室
24	Conference Room	24	大会议室
25	Staff Reading Room	25	员工阅览室

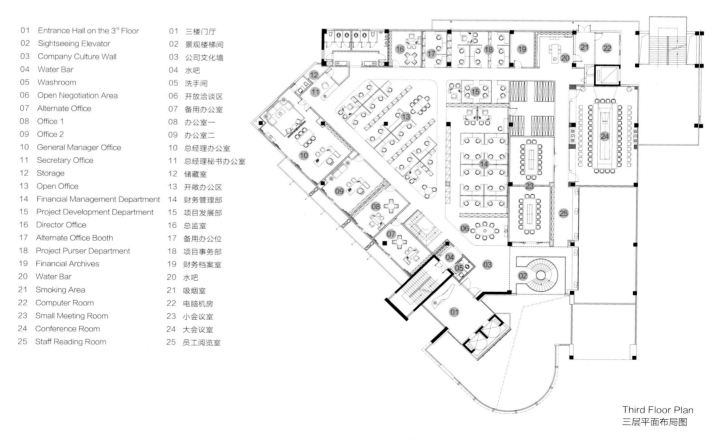

Third Floor Plan
三层平面布局图

01	Reception Area	01	接待区
02	Foyer	02	玄关
03	Lounge	03	休息区
04	Tea Room	04	茶水间
05	Meeting Room	05	会议室
06	Multifunction Hall	06	多功能厅
07	Equipment Room	07	设备房
08	Video Conference Room	08	视频会议室
09	Reception Room	09	接待室
10	Waterscape	10	水景
11	Public Washroom	11	公共卫生间
12	Alternate Office	12	备用办公室
13	Office 1	13	办公室一
14	Office 2	14	办公室二
15	Open Workspace	15	开敞办公区
16	Secretary-General Office	16	总秘处
17	General Manager Office	17	总经理办公室
18	Manager Workstation	18	经理位
19	Office 3	19	办公室三
20	Office 4	20	办公室四
21	Alternate Office 2	21	备用办公室二
22	Smoking Room	22	吸烟室
23	Computer Room	23	电脑机房
24	Corridor	24	走道
25	Design Dedicated Storeroom	25	设计专用库房
26	Alternate Storage	26	备用库房
27	Lounge	27	休息区
28	Small Meeting Room	28	小会议室

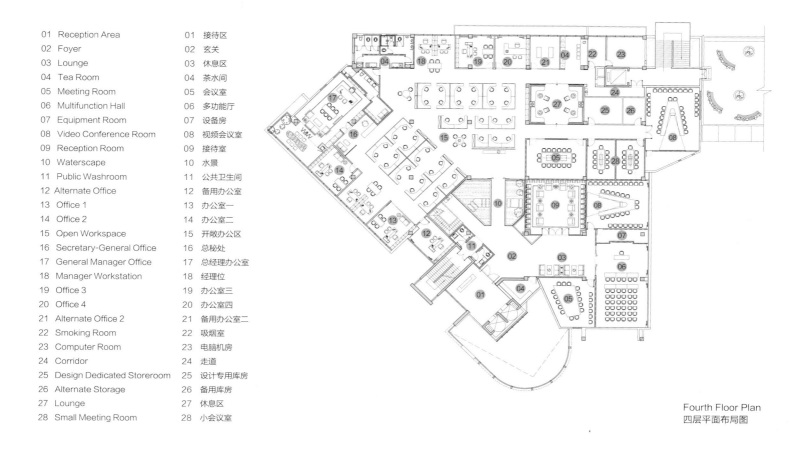

Fourth Floor Plan
四层平面布局图

成都万科金域缇香销售中心

成都 / 中国

韵律的节拍

设计师运用干净凝练的线条勾勒了整个空间格局，进行排比铺设，呈现出的韵律与气质好似古典乐曲的节奏感及感染力。材料的运用繁简相宜，石材丰富的天然纹理，搭配水曲柳温润的木纹，在软装表现上，圆形的吊灯打破空间的刚硬，在家私上更是刚中带柔，营造舒适开放自由的氛围。

成都万科金域缇香销售中心荣获2014年德国IF大奖

CHENGDU VANKE JINYU TIXIANG SALES CENTER

CHENGDU / CHINA

Beat of Rhythm

The designer constructed the entire space structure with clean and concise lines arranging in rows; the rhythm and temperament it presents is just like a classical music. Natural grains of stone are combined with mild texture of ash tree; materials used here are controlled in a balance between magnificent and simple. In terms of soft decoration, a round pendant light neutralizes the crudeness of the space, and the furniture is all selected in a balance between hardness and softness to create an open free atmosphere.

Chengdu Vanke Jinyu Tixiang Sales Center got Excellence of 2014 German IF award

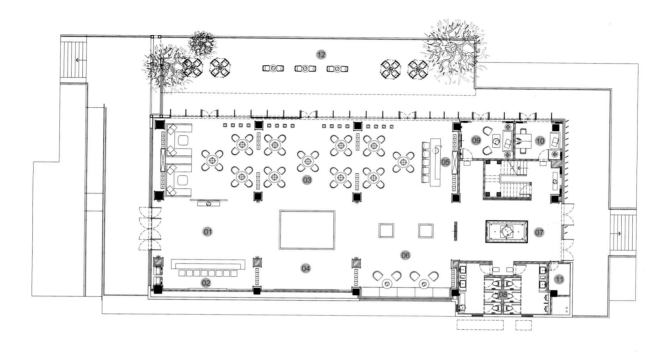

Ground Floor Plan
平面布局图

01 Entrance Hall   01 前厅
02 Reception Desk   02 接待前台
03 Discussion Area   03 洽谈区
04 Sand Table Area   04 沙盘区
05 Water Bar   05 水吧区
06 Special Discussion Area   06 特色洽谈区
07 Waterscape   07 水景
08 Men & Women's Washroom   08 男女卫生间
09 VIP Room   09 VIP室
10 Office   10 办公室
11 Computer Room   11 机房
12 Outdoor Landscape Area   12 户外景观区

贵阳俊发办公样板房

贵阳 / 中国

空间是由黑色的构架结合功能勾勒出建筑感的轮廓，起起落落，弯弯折折，串联在空间里。那自然又该如何解读呢？设计师运用色彩进行了一些模拟，深灰色的地毯是土壤，绿色是小草，黄色是油菜花及向日葵，蓝色是天空，橡木饰面是绿草、黄花、蓝天、土壤下的小木屋。

贵阳俊发办公样板房荣获美国国际室内设计协会大奖
——卓越奖

GUIYANG JUNFA OFFICE SAMPLE HOUSE

GUIYANG / CHINA

The outline structure of the whole space is formed by black frameworks combining with multi-functions, those ups and downs, curves and folds, connecting in series in the space. Then how is nature being reflected? Colors are used in the simulation, dark gray carpet as soil, green as grass, yellow as rape flower or sun flower, blue as sky and oak veneers as log cabins among the green grass, yellow flower, blue sky and dark gray land.

Guiyang Junfa Office Sample House got Excellence Prize of American IIDA award

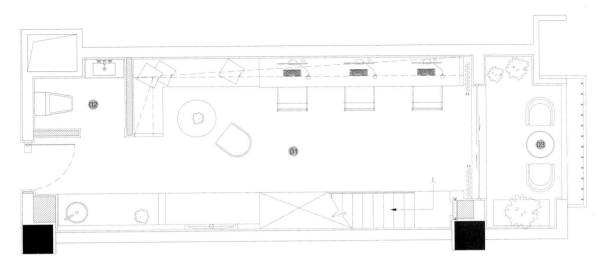

01 Office   01 办公室
02 Washroom   02 卫生间
03 Balcony   03 阳台

First Floor Plan
一层平面布局图

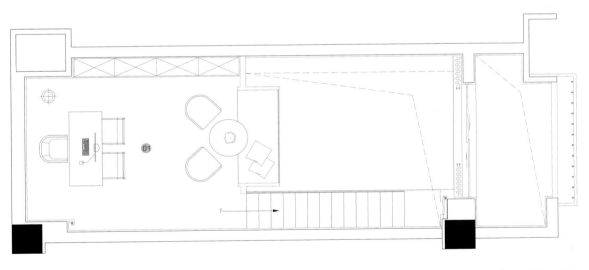

01 Loft Office   01 阁楼办公室

Second Floor Plan
二层平面布局图

QUANZHOU WANDAO ZIYUNTAI SALES CLUB

QUANZHOU / CHINA

泉州万道紫云台销售会所

泉州 / 中国

Model for creative inheritance of classical architecture

This project re-interpreted new elements and inspirations obtained from traditional folk houses in Quanzhou. Courtyard structure is with a history of millennium years, and inside the space there listed the worldwide famous Dehua Ceramic Whiteware, Hui'an Stone Carving, Xianyou Wood Carving; apart from all the above, this is more of an epitome of humanity, nature and history. The combination and contrast between dynamic and static, more and less, new and old, straight and curve, sky and earth, human and space, all formed the "NEW ASIA" space with both inheritance and breakthroughs.

古典新传承的典范

本项目从泉州传统民居中提炼新的元素与灵感，进行重新解读。院落的格局延续了千年，空间中包含了享誉海内外的德化白瓷、惠安石雕、仙游木雕，这里更有人文的、自然的、历史的缩影。空间中的动与静、多与少、新与旧、直与曲、天与地、人与空间的组合与对比，形成了既有传承又有突破的时尚"新亚洲"空间。

泉州万道紫云台销售会所荣获美国国际室内设计协会大奖——卓越奖

Quanzhou Wandao Ziyuntai Sales Club got Excellence Prize of American IIDA award

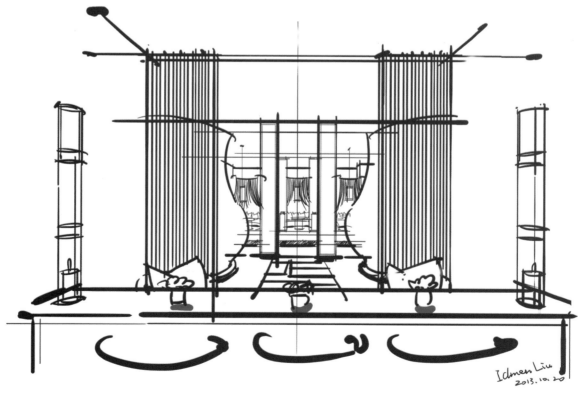

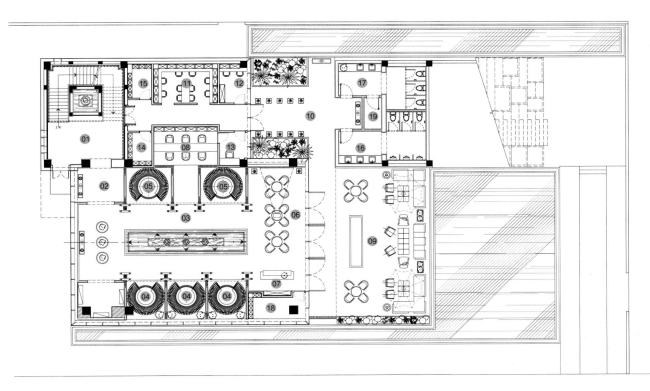

First Floor Plan
一层平面布局图

01	Panoramic Stairwell	01	景观楼梯间
02	Hallway	02	过厅
03	Waterscape	03	水景
04	Depth Discussion Room	04	深度洽谈区
05	VIP room	05	VIP室
06	Audio-Visual Area	06	影音区
07	Water Bar	07	水吧
08	Contract Area	08	签约区
09	Landscape Terrace	09	景观露台
10	Hallway	10	过厅
11	Separate Office	11	独立办公室
12	Property Management Office	12	物业管理办公室
13	Financial Room	13	财务室
14	Men's Dressing Room	14	男更衣室
15	Women's Dressing Room	15	女更衣室
16	Men's Room	16	男卫
17	Women's Room	17	女卫
18	Storage	18	储藏室
19	Cleaning Room	19	保洁室

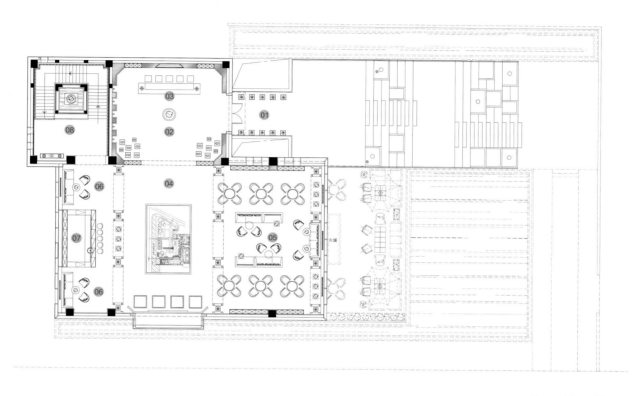

Second Floor Plan
二层平面布局图

01 Art Gallery
02 Reception Desk
03 Sales Control Desk
04 Model Show Area
05 Discussion Area
06 Lounge
07 Water Bar
08 Stairwell

01 艺术过廊
02 前台接待
03 销控台
04 模型区
05 洽谈区
06 休闲区
07 水吧
08 电梯间

泉州万道办公室

泉州 / 中国

根据业主要求把办公室会所化,营造一种让员工轻松愉快地工作的氛围。大致的空间属性分为两种,休息区和水吧及贵宾接待室相对灯光氛围比较强烈;工作区及讨论区相对敞亮开阔。不同属性的材料质感相互对比、呼应,在空间中融入一些风情的色彩,搭配艺术化的装饰品,充分体现了业主及企业的品位倾向。设计师在把握功能性与艺术性相互结合这个"火候"上,控制得恰到好处。

QUANZHOU WONDERS OFFICE

QUANZHOU / CHINA

Respecting to owner's requirements, the office is designed with club style to create delightful working atmosphere for the staff. Generally there are two space types, lighting atmosphere in rest room, water bar and VIP reception room is relatively of strong contrast and special lighting effect; while working area and meeting room are relatively broad and bright. Different materials are in silent contrast and connection; few special style colors and artistic decorations fully indicate the owner's taste and preferences. Designers handle the combination of functionality and artistry in a very subtle way; every part is within rather appropriate control.

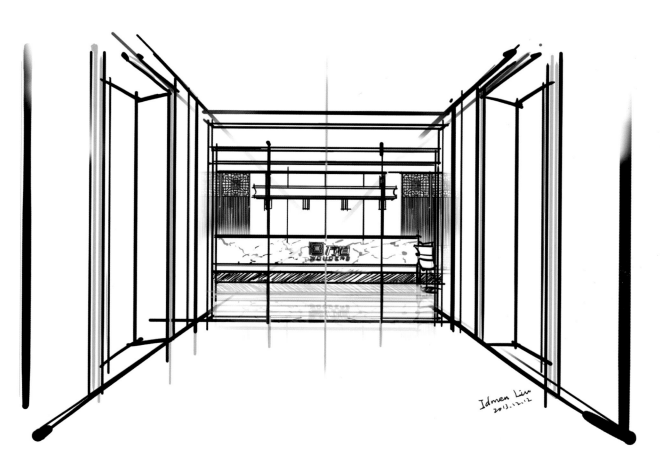

Ground Floor Plan
平面布局图

01	Elevator Waiting Hall	01	电梯厅	18	Open Workspace	18	开敞办公区
02	Reception Desk	02	前台	19	Reception Area	19	接待区
03	Discussion Room	03	洽谈室	20	VIP Room	20	贵宾室
04	Diver's Lounge	04	司机休息室	21	Reading Area	21	阅读区
05	Equipment Room	05	设备间	22	Entertaining Area	22	娱乐区
06	Central Air-Conditioning Room	06	空调主机位置	23	Water Bar	23	水吧区
07	Large Conference Room	07	大会议室	24	Storage	24	储藏室
08	Meeting Room	08	会议室	25	Material Storage Room	25	材料储藏室
09	Financial Room	09	财务室	26	Reference Room	26	资料储藏室
10	Archive Room	10	档案室	27	Centre General Office 1	27	中心总办公室一
11	Vice-President Room 1	11	副总裁室二	28	Centre General Office 2	28	中心总办公室二
12	Boardroom	12	董事室	29	Centre General Office 3	29	中心总办公室三
13	Vice-President Room 2	13	副总裁室一	30	Centre General Office 4	30	中心总办公室四
14	President Office	14	总裁室	31	Printing Area	31	文印区
15	Chairman Secretary Office	15	董秘办公室	32	Open Discussion Area	32	开放讨论区
16	Lounge	16	休息室	33	Centre General Office 5	33	中心总办公室六
17	Board Chairman Office	17	董事长室	34	Centre General Office 6	34	中心总办公室五

重庆万科城别墅

重庆 / 中国

本案在新亚洲韵味中，融入法式的浪漫情怀。在淡雅大方的硬装环境下，设计师采用对称、整列的手法，旨在糅合中西、升华意境。干练有力的线条，点缀些许精致的欧式纹案；灰色主调在冷蓝色的调和下，同时跳跃点滴的暖色，显得格外沉稳、利落；在细节处，暗调的纹理值得关注。在整体硬朗的环境下，配合软装的温暖、质感，整个空间显得格外耐人寻味，可堪品味。

CHONGQING VANKE TOWN VILLA

CHONGQING / CHINA

French romance is mixed with new Asian style in this case. Designers use symmetry and array in the simple and elegant hard environment to achieve the integration of Chinese and western style and the sublimation of artistic taste. Clear and strong lines are embellished with few European style texture; cold blue interacting with grey main color, and with dots of warm color jumping in, appears particularly stable and vivid; dark color textures are worth of attention for details. The entire space is rather enchanting and worth of appreciation in the solemn environment along with warm feeling and quality sense of soft decorations.

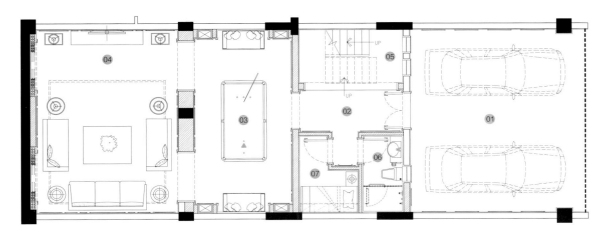

B1 Floor Plan
负一层平面布局图

01　Garage
02　Entrance Hall
03　Billiard Room
04　Audio-Visual Room
05　Stairwell
06　Washroom
07　Housemaid Room

01　车库
02　玄关
03　桌球室
04　视听室
05　楼梯间
06　洗手间
07　保姆房

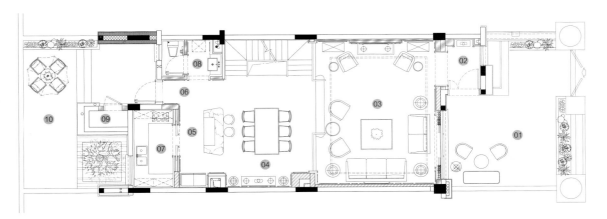

First Floor Plan
一层平面布局图

01　Courtyard
02　Entrance Hall
03　Living Room
04　Dining Room
05　Bar
06　Hallway
07　Kitchen
08　Washroom
09　Storage
10　Terrace

01　庭院
02　玄关
03　客厅
04　餐厅
05　吧台区
06　过厅
07　厨房
08　洗手间
09　储物间
10　露台

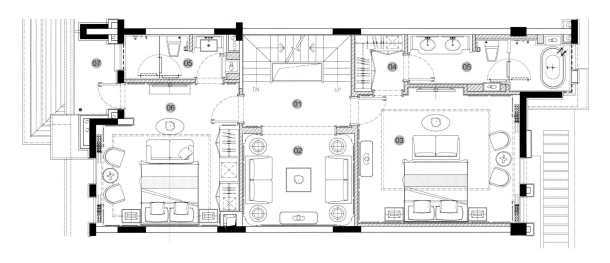

Second Floor Plan
二层平面布局图

01	Entrance Hall	01	过厅
02	living Room	02	起居室
03	Bedroom 1	03	卧室一
04	Walk-in Closet	04	衣帽间
05	Washroom	05	洗手间
06	Bedroom 2	06	卧室二
07	Relaxing Terrace	07	休闲阳台

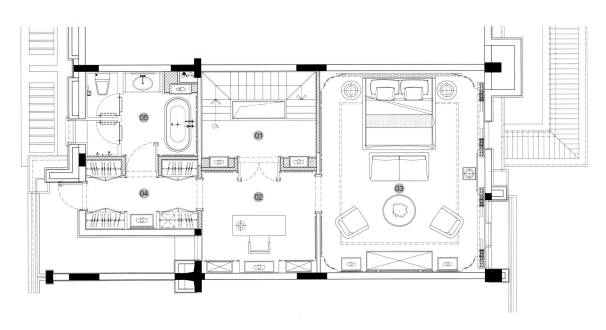

Third Floor Plan
三层平面布局图

01	Hallway	01	过厅
02	Study	02	书房
03	Master Bedroom	03	主卧室
04	Walk-in Closet	04	衣帽间
05	Washroom	05	洗手间

路劲集团（常州）X1户型样板房

常州 / 中国

ROAD KING GROUP X1 TYPE SAMPLE HOUSE, CHANGZHOU

CHANGZHOU / CHINA

空间设计中元素符号的运用为整个空间增添了一丝东方美学的气息，配以一丝不苟的艺术品，柔和妩媚的灯光效果和丝绸、棉麻的针织布艺，充分突显了设计师对装饰艺术风格及典雅居住环境的不懈追求。

The element symbol used in the space design added for the space a feeling of oriental aesthetics, and along with the delicate art pieces, soft lighting effect, knitting fabrics of silk and cotton linen, it adequately reflected the decorative art style and the designers' ceaseless pursuit for elegance.

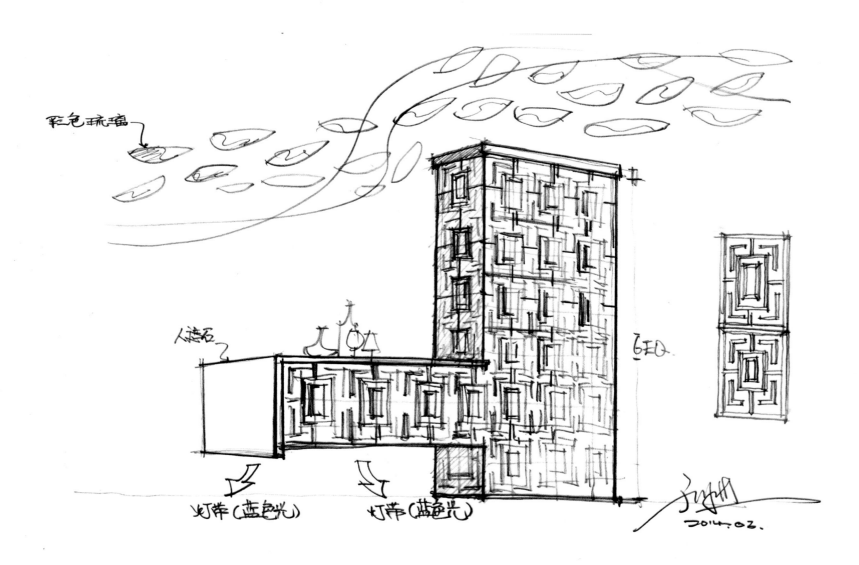

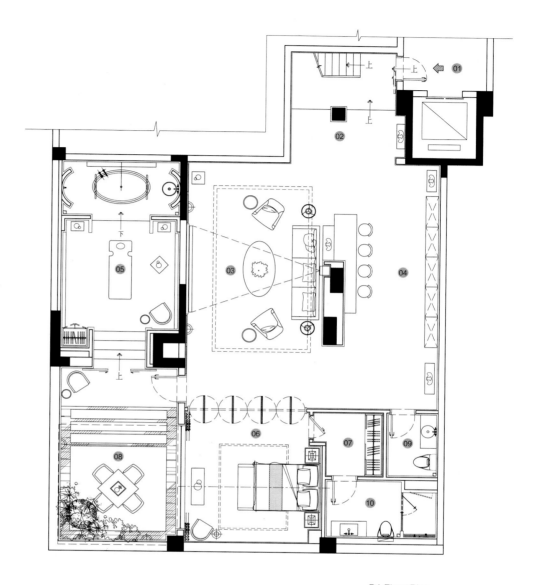

B1 Floor Plan
负一层平面布局图

01	Garage Entrance	01	车库入口
02	Hallway	02	过厅
03	Cinema	03	影音区
04	Water Bar	04	水吧
05	Spa Room	05	SPA房
06	Guestroom	06	客房
07	Walk-in Closet	07	衣帽间
08	Entertainment Terrace	08	休闲露台
09	Washroom	09	卫生间
10	Guestroom Bathroom	10	客卫

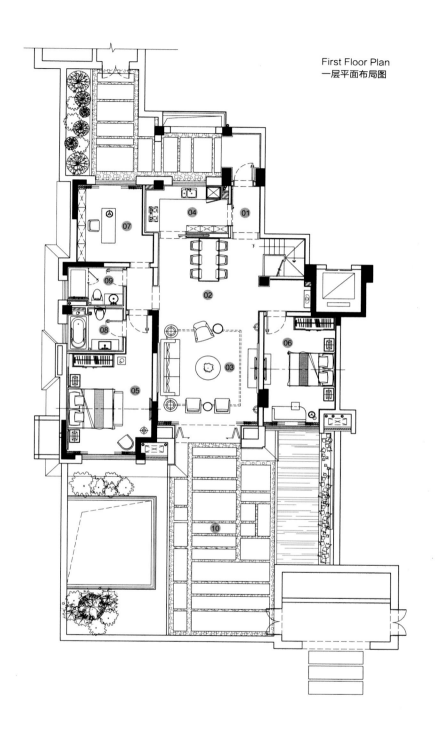

01	Hallway
02	Dining Room
03	Living Room
04	Kitchen
05	Master Bedroom
06	Second Bedroom
07	Study
08	Main Bathroom
09	Washroom
10	Courtyard

01	玄关
02	餐厅
03	客厅
04	厨房
05	主卧
06	次卧
07	书房
08	主卫
09	卫生间
10	庭院

First Floor Plan
一层平面布局图

ROAD KING GROUP X2 TYPE SAMPLE HOUSE, CHANGZHOU

CHANGZHOU / CHINA

路劲集团（常州）X2户型样板房

常州 / 中国

本设计不局限于形式上的奢华，而更趋向于一种不同材质之间的搭配运用，并着重于个性、时尚，去繁求简，珍视功能及自然美学。各个空间无不流露着大自然的气息。设计师以"个性化"的表达方式，展现超越现实的想象空间。同时也注重功能性、舒适性、空间形态与色彩体块的统一性。墙面以原木色突出设计形态，用热情奔放的设计语言描绘现代、个性、时尚的高品质生活。

This design case does not limit itself in modality luxury but promotes to the match of different materials, and it pays much attention to personality, fashion, function and natural aesthetics. Breath of nature is everywhere in the space. Designers use "personalized" method to show us an imagined space beyond reality. At the mean time, functions, comfort and unity between space forms and color zones have been paid much importance to here. Burlywood wall highlights the primitive design form, which is how designers describe the modern, personalized, fashionable quality life with passionate design language.

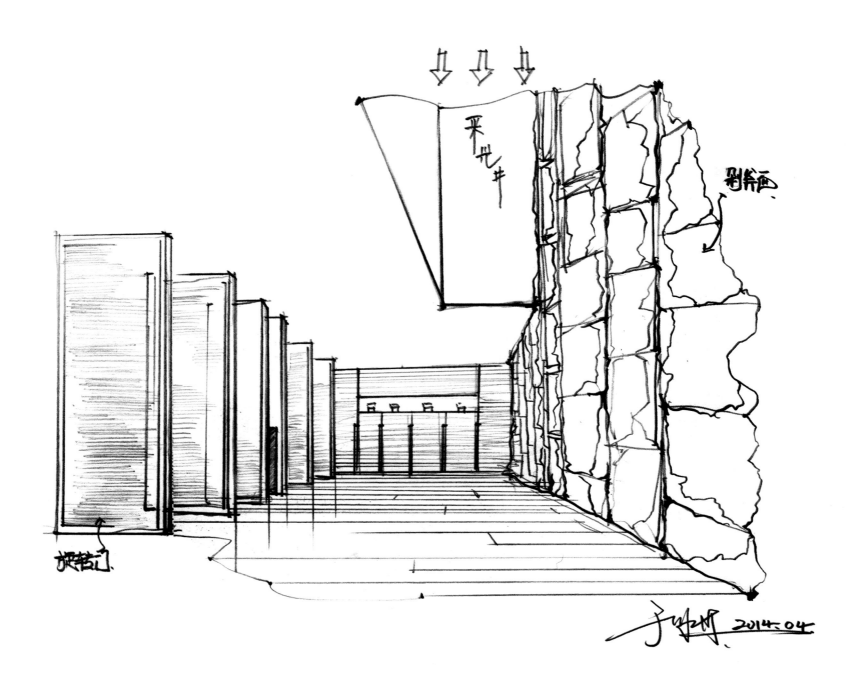

01 Audio-Visual Room
02 Water Bar
03 Drawing Room
04 Cellar
05 Washroom

01 视听室
02 水吧
03 书画室
04 酒窖
05 洗手间

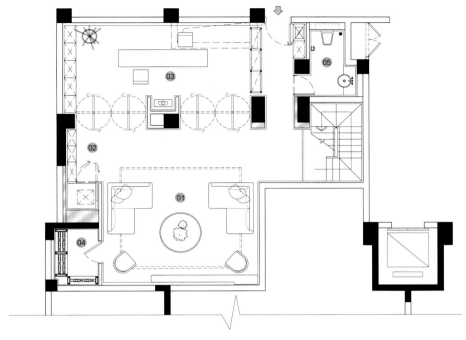

B1 Floor Plan
负一层平面布局图

01 Hallway
02 Dining Room
03 Living Room
04 Kitchen
05 Master Bedroom
06 Second Bedroom
07 Children's Room
08 Main Bathroom
09 Washroom
10 Terrace
11 Balcony

01 玄关
02 餐厅
03 客厅
04 厨房
05 主卧
06 次卧
07 儿童房
08 主卫
09 卫生间
10 露台
11 阳台

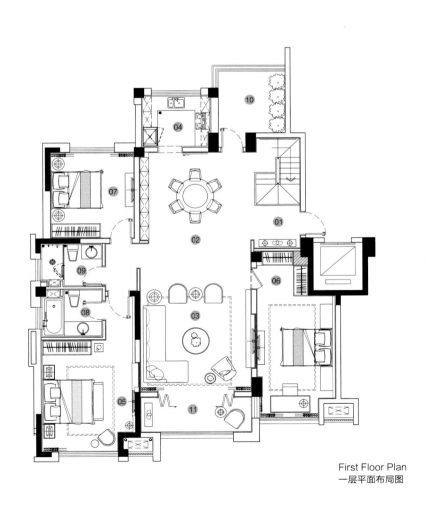

First Floor Plan
一层平面布局图

路劲集团（常州）Y1户型样板房

常州 / 中国

ROAD KING GROUP Y1 TYPE SAMPLE HOUSE, CHANGZHOU

CHANGZHOU / CHINA

整个设计在沿用古典欧式的传统基础上，摒弃繁复，不失简约、高贵、时尚。整个空间布局进行分割处理，使空间自然流畅、井然有序，并透露出高贵与奢华。在规划布局上营造出浓郁的欧洲风格。细节的设计充分尊重自然、拥抱健康，低调中流露出欧洲的浪漫主义风情。因此，在设计上不但要考虑建筑内外的结合，更要注重考虑项目本身的高贵品质。整体暖灰色空间搭配色彩跳跃的软装元素来打破传统格调，赋予整个空间时尚感，营造出迷人的、充满艺术气质的氛围。

On the basis of Classical & European style, designers abandon complexity and keep its simplicity, nobility and fashion. The entire space is naturally separated into well-regulated sections of pure European style. Design of details presents sense of nature and health, fully expressing the romance of European style. In contrast to the main color warm grey, bright colors of soft decorations break the heaviness of traditional style and put fashion sense into the space, creating a charming and artistic temperament. Therefore, designers should not only pay attention to the combination of exterior and interior of architecture, but more importantly should consider to the noble quality of the project.

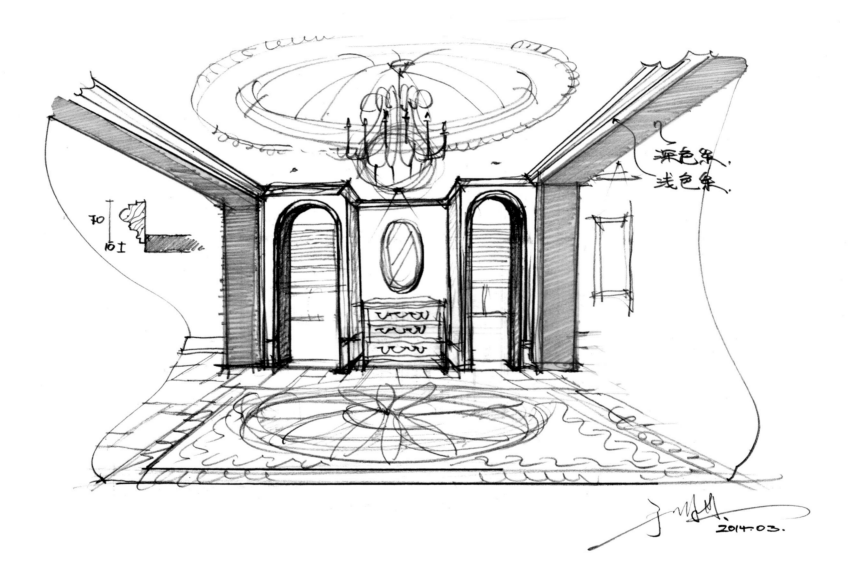

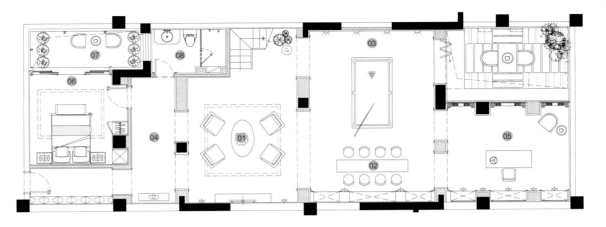

B1 Floor Plan
负一层平面布局图

01	Family Bar	01	家庭吧
02	Water Bar	02	水吧台
03	Entertainment Area	03	娱乐区
04	Hallway	04	过厅
05	Study	05	书房
06	Guestroom	06	客房
07	Sunk Courtyard	07	下沉庭院
08	Washroom	08	洗手间

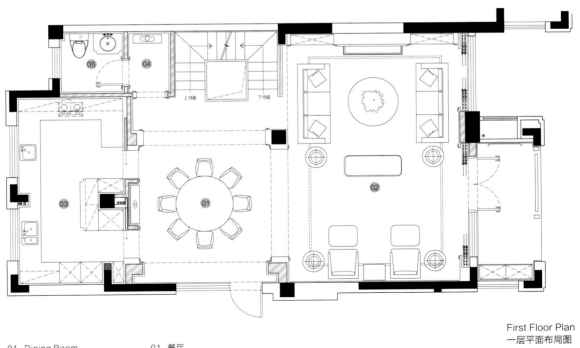

First Floor Plan
一层平面布局图

01	Dining Room	01	餐厅
02	Guestroom	02	客厅
03	Kitchen	03	厨房
04	Hallway	04	玄关
05	Washroom	05	洗手间

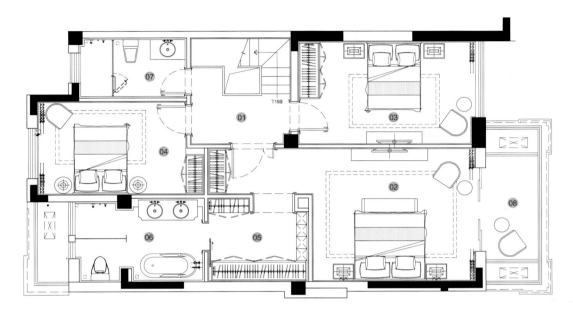

Second Floor Plan
二层平面布局图

01　Hallway　　　　　01　过厅
02　Master Bedroom　 02　主卧
03　Second Bedroom　 03　次卧
04　Children's Room　04　小孩房
05　Cloakroom　　　　05　衣帽间
06　Main Bathroom　　06　主卫
07　Washroom　　　　 07　卫生间
08　Terrace　　　　　08　露台

重庆万科悦湾别墅

重庆 / 中国

CHONGQING VANKE YUE BAY VILLA

CHONGQING / CHINA

设计师将传统中式元素与东南亚风情结合运用，平面采用隐约的对称布局，端正稳健，充分体现出中国传统精神美学。立面运用现代设计手法将东方的语汇去繁从简表达出来。纯粹的橡木染色饰面，浅色的布料、墙纸，格调高雅，造型简朴优美，色彩对比浓郁而成熟，营造出中国传统中式的庄重与东南亚风情的优雅气质。

Designers combined traditional Chinese elements with Southeast Asian Style and adopted indistinct symmetric structure for planar layout which is dignified and modest, fully showing Chinese traditional spiritual aesthetic. For facade, designers presented oriental glossary with modern design method in which way simplicity remained and redundancy abandoned. Pure oak dyed veneer, light color cloth and wall paper are elegant, simple and beautiful, forming strong and mature color contrast. All these represent the Chinese antique solemnity and Southeast Asian elegance.

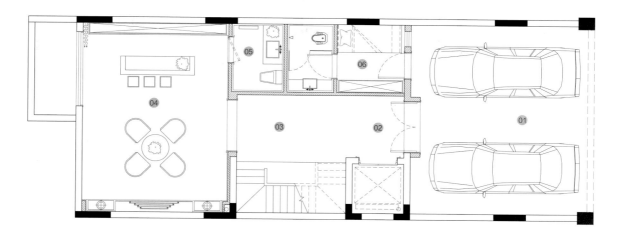

B2 Floor Plan
负二层平面布局图

01	Garage	01	车库
02	Entrance Hall	02	玄关
03	Hallway	03	过厅
04	Red Wine & Cigar Bar	04	红酒雪茄吧
05	Public Washroom	05	公卫
06	Staff Room	06	工人房

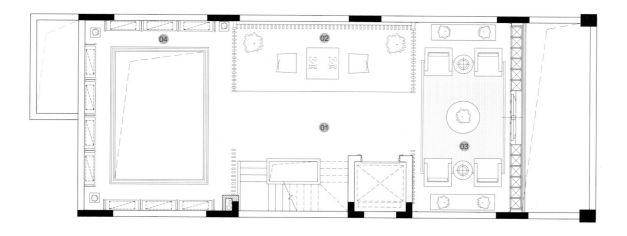

B1 Floor Plan
负一层平面布局图

01	Hallway	01	过厅
02	Tea Taste Area	02	品茶区
03	Family Room	03	家庭厅
04	Art Gallery	04	收藏艺术回廊

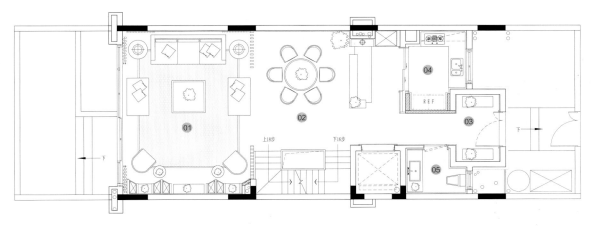

First Floor Plan
一层平面布局图

01 Living Room 01 客厅
02 Dining Room 02 餐厅
03 Entrance Hall 03 玄关
04 Kitchen 04 厨房
05 Washroom 05 洗手间

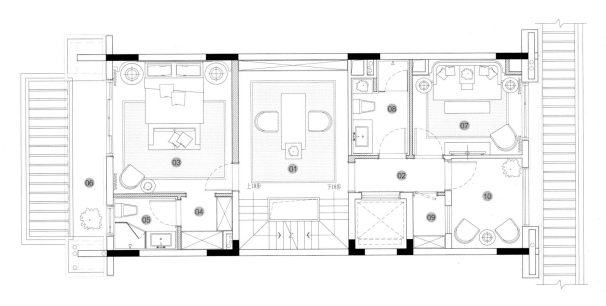

Second Floor Plan
二层平面布局图

01 Study 01 书房
02 Hallway 02 过厅
03 Second Bedroom 03 次卧室
04 Cloakroom 04 衣帽间
05 Second Bathroom 05 次卫
06 Terrace 06 露台
07 Multifunction Room 07 多功能房
08 Washroom 08 洗手间
09 Storage 09 贮藏室
10 Landscape Balcony 10 景观阳台

301

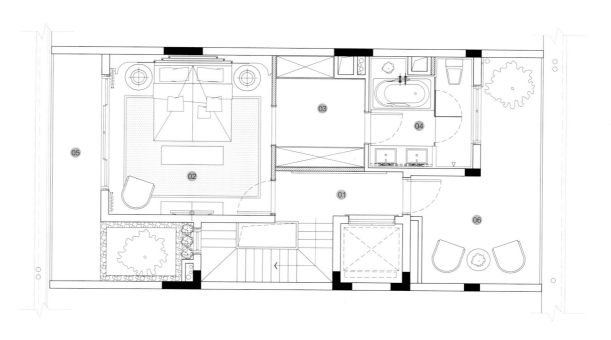

Third Floor Plan
三层平面布局图

01 Hallway 　　　　01 过厅
02 Master Bedroom 　02 主卧室
03 Cloakroom 　　　 03 衣帽间
04 Main Bathroom 　 04 主卫
05 Terrace 　　　　 05 露台
06 Landscape Balcony 06 景观阳台

深圳彭成大东城别墅

深圳 / 中国

1. 风格定义

法式混搭风格融合工业文化和欧洲装饰独有的特色，以宽大、舒适、杂糅而著称。

2. 空间分析

一层南北通透采光极佳，配合户外庭院相得益彰，近7米高的客厅空间和宽敞方正的餐厅空间以及开放式厨房无不体现出法式风格宽大舒适的特点。二层女儿房运用蓝色基调，煽情摆件勾画出少女情怀。三层为主人的私有空间，卧室都基本保留了原有层高，空间方正大气。家具饰品延续混搭风格特色。主人房主卫功能齐备，天窗采光设计为主人的生活添加情趣的同时也延展了空间。独立的衣帽间配有临窗梳妆台方便实用。地下一层为家庭娱乐区，配备影音室、家庭娱乐室，在两个空间中间设置了豪华气派的旋转楼梯，通往地下二层的楼梯气势磅礴，加深了对豪宅的定义。地下二层由主人收藏空间和佣人功能操作空间两部分构成，安静惬意的环境更好地营造出收藏室的神秘感。佣人房间和操作区靠近车库入口，使主仆分区方便使用。

3. 家具饰品

众所周知，欧洲人在对待家具的方面是本着"愈久愈好"的心态的，这根源于其历史文化，法式混搭风格继承和发展了欧式家具的这一传统，并且在强调巴洛克和洛可可的浮华与新奇的同时又加入了工业化时尚家具，产生极强的视觉冲击。工业化时尚家具强调简洁、明晰的线条和复古做旧的装饰。家具色彩大多会加上金属配件或其他粗犷的装饰，突出其风格个性。总之各个空间均恰当地将家具摆放其中，营造宽大、舒适、杂糅的法式混搭风格。

SHENZHEN PENGCHENG DADONGCHENG VILLA

SHENZHEN / CHINA

1. Style Definition

French fusion. Mix match of French style, industrial culture and European decorations. Featured by spacious, comfortable and blending.

2. Space Analysis

The first floor, fully connected from South to North, perfect day-lighting, with outdoor garden, living room of 7 meters high, square and broad dining room, and open kitchen, all embody French style's feature of spaciousness and comfort. The second floor, daughter room, Whose basic color is blue, delicately displays furniture full of maiden complex. The third floor is owner's private space, with bedroom of high ceiling, and square and broad space. Furniture is still in mix match style. The owner's bedroom is equipped with full functional main bathroom; the skylight extends space and creates delight of life. Independent cloakroom has dresser near the window. The first floor underground is family entertainment area. Audio-video room and entertaining room are separated by luxurious spiral stair down to the second floor underground, which highlights the definition of luxury villa. The second floor underground falls into owner's collecting space and servants' working space; very quiet and cozy environment creates a sense of mystery for the collecting space. Servants' room and working space are close to garage, which is convenient for independent activities.

3. Furniture and Decoration

As everyone knows, Europeans have a feeling for furniture of "long history" which originates from their historic culture. French fusion style inherits and develops this tradition; what's more, it adds in industrial fashion while emphasizing Baroque and Rococo. Industrial fashion furniture attaches importance to simplicity, clear lines and vintage faded decorations. The furniture colors are usually added with metal fixtures or other rough fixtures to highlight its characteristic. All in all, every space is set with appropriate furniture and decorations to build the spacious, comfortable and blending French fusion style.

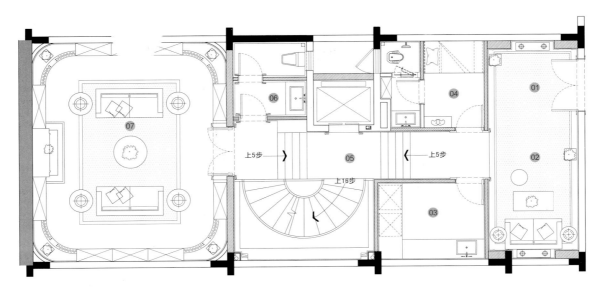

B2 Floor Plan
负二层平面布局图

01 Indoor Hallway　　　　　　01 内玄关
02 Entrance Hall　　　　　　　02 入户大堂
03 Workspace　　　　　　　　03 工作间
04 Staff Room　　　　　　　　04 工人房
05 Staircase　　　　　　　　　05 楼梯间
06 Washroom　　　　　　　　 06 洗手间
07 Family History Memorial Room　07 家族历史纪念馆

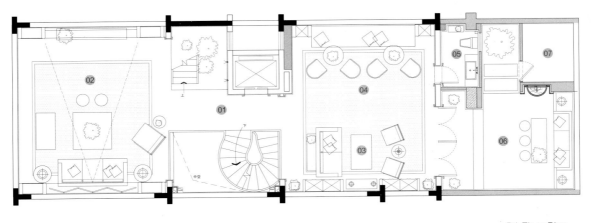

B1 Floor Plan
负一层平面布局图

01 Corridor　　　　　　　　　01 过道
02 Audio-Visual Room　　　　 02 影音室
03 Chess & Card Game Room　03 棋牌游戏区
04 Tea Area　　　　　　　　　04 品茶区
05 Washroom　　　　　　　　 05 洗手间
06 Family-Gather BBQ　　　　 06 家庭聚会BBQ
07 Storage　　　　　　　　　 07 杂物间

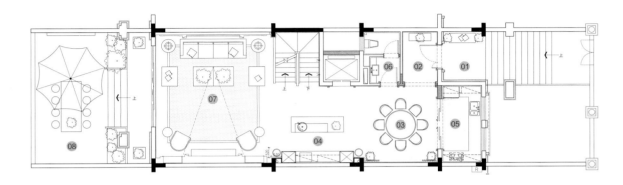

First Floor Plan
一层平面布局图

01 Entrance Hall   01 玄关
02 Interior Hall   02 内玄关
03 Dining Room   03 餐厅
04 Water Bar   04 水吧
05 Kitchen   05 厨房
06 Washroom   06 洗手间
07 Living Room   07 客厅
08 Outdoor Courtyard   08 户外庭院

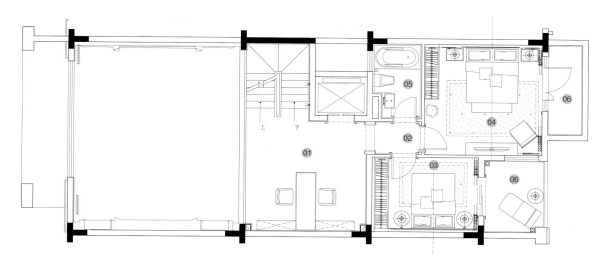

01	Family Room	01	家庭厅
02	Hallway	02	玄关
03	Boy's Bedroom	03	男孩房
04	Girl's Bedroom	04	女孩房
05	Washroom	05	洗手间
06	Landscape Balcony	06	景观阳台

Second Floor Plan
二层平面布局图

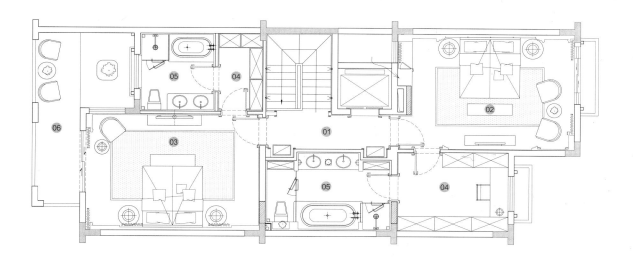

01	Hallway	01	玄关
02	Mater Bedroom	02	主人房
03	The Elder's Bedroom	03	老人房
04	Walk-in Closet	04	衣帽间
05	Washroom	05	卫生间
06	Balcony	06	阳台

Third Floor Plan
三层平面布局图

深圳臻络科技办公室

深圳 / 中国

没错！您看到的是一个深圳优秀的新兴科技公司办公环境，设计初期和GYENNO创始人任康的交流特别简单，他说："我需要一个区别于以往的科技公司的办公环境"。仅此而已的设计要求，这是基于对设计师的充分信任和支持。摆在设计师面前的有两个难点：第一个难点是设计造价，一般初期创业公司在办公室空间投入上肯定会比较谨慎，这也正是此项目非常有挑战性的一点；第二个难点就是风格，怎么改变我们对科技公司的一般印象？应该如何去营造一个不一样的办公氛围？真正好的设计不是炫技，也不是靠高投入才能高产出，设计应该是用心用情去对待每个项目。废话不多说，请看完成之后的现场照片。

SHENZHEN GYENNO TECHNOLOGY OFFICE

SHENZHEN / CHINA

Yes! This is the office of an outstanding new-rising technology company in Shenzhen. The communication with GYENNO founder Ren Kang about the design is very simple, who said, "I want an office different from technology companies in common sense". This unique design requirement is based on his full trust and support on the designers. There are two difficulties for designers: first is design cost, a starter company is usually careful on office cost which is a great challenge for the project; second is style, how to change the common impression towards technology company? What should be done to build a special office environment? The best design is not based on a showing of design methods, neither on high cost. It is your full heart devotion to make it the best. Here are the photos on site.

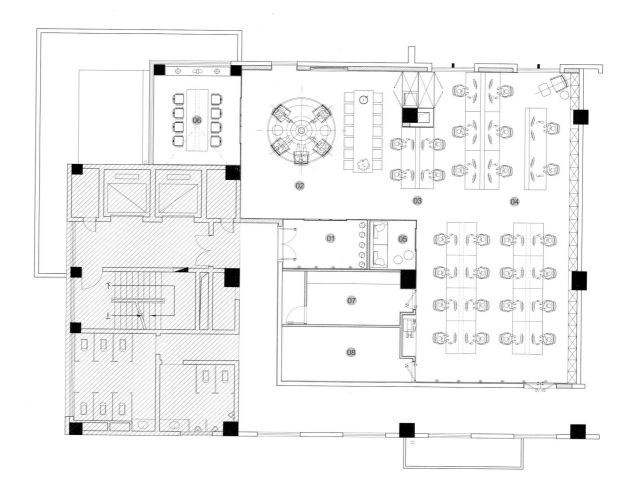

Ground Floor Plan
平面布局图

01 Entrance Hall   01 入口前厅
02 Discussion Area   02 讨论区
03 Transitional Area   03 临时区
04 Workspace   04 工作区
05 Lounge   05 休息室
06 Meeting Room   06 会议室
07 Lab   07 实验室
08 Products Storage   08 产品储藏室

济南路劲销售中心

济南 / 中国

JI'NAN ROAD KING SALES CENTER

JI'NAN / CHINA

本楼盘主要面向时尚、舒适、精致、阳光、热爱生活的年轻客户群体，不需要过多的华丽修饰，而更注重突出空间的功能性、实用性、美观感，呈现空间的多维性。

本案从建筑到室内，整个空间就像是一个巨大的盒子。我们试图运用构成中的点、线、面关系，来搭建一个简洁、时尚的功能空间。运用暖暖的米色系材料使得空间更加富有亲和力与精致感。

地面引导性的石材铺贴、墙体的形式造型、呈现空间的不同组合的线条，增加了室内的引导性与空间感。

空间造型上，以模型展示区与客户洽谈区两个大型造型天花将空间划分。中间的户型模型以天花悬挂形式呈现，将空间划分，各自独立却又因为它的存在而巧妙衔接在一起。空间被赋予更多整体感与灵动性。

因为建筑周边的景观原因，大型的玻璃幕墙外并没有太多的美观感。考虑到这一点，设计师通过运用铝合金格栅错落地竖向悬挂在天花板上，既有连续性又有层次感，从视觉上形成一幅整体又富有连贯性的山水画，从而体现一种舒适、环保、乐活的生态居住理念。

This project is designed for young people in pursuit of fashion, comfort, delicacy, sunshine and passion for life. Therefore there is no need of too much luxuriant decorations; more attention is attached to function, utility and space structure.

From architecture structure to interior environment, the entire space is like a huge box. We try to use the relationship between its points, lines, and areas to build a simple and fashion functional space. Cream-color materials make the space more cozy and attractive.

Introductory marble floor, special wall structure and different space lines make the house more directional and dimensional.

On space arrangement, the whole space is separated by the two large ceiling of model display area and customer negotiation area. Models in the middle area are displayed by hanging ceiling. In this case, different areas are flexibly well-organized, independent yet connected.

Views out of the large glass walls are not so attractive due to the existing surroundings. Considering this issue, designers build aluminum grids hanging down from ceiling in different lengths, which creates a view of landscape painting. And that is the ecological living concept of comfort, environmental protection and happy living.

333

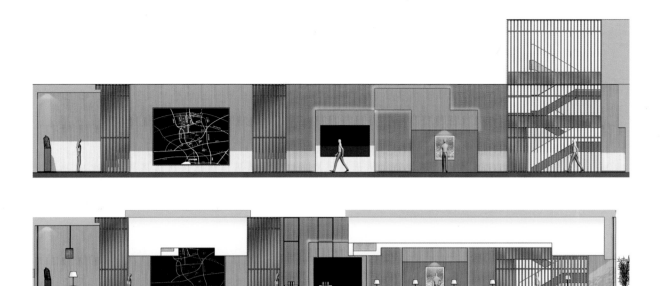

01 Reception Area
02 Sand Table Area
03 Condo Model Show Area
04 Discussion Area
05 Depth Discussion Area
06 Water Bar
07 Online Contract Area
08 Children Amusement Area
09 Contract Area
10 Financial Area
11 Sales Management Room
12 Washroom

01 接待区
02 沙盘区
03 分户模型
04 洽谈区
05 深入洽谈区
06 水吧区
07 网签区
08 儿童娱乐区
09 签约区
10 财务
11 销管
12 洗手间

Ground Floor Plan
平面布局图

ROAD KING GROUP X6 TYPE SAMPLE HOUSE, CHANGZHOU

CHANGZHOU / CHINA

路劲集团（常州）X6户型样板房

常州 / 中国

本案以都市黑白的简约时尚设计风格为基调，以米色系的橡木作为空间的温和调色剂，构筑一个现代简约、时尚个性又不失温馨、优雅的空间。

从步入空间的客厅、餐厅到小孩房、主卧房，设计师希望在沿袭现代简约风格的基础上，使得空间更加有主人自己的个性与喜好。餐厅区域的一幅都市黑白照片墙的出现，彰显了空间的个性与艺术感。卧房床头的挂饰体现了设计师对生活的幽默感，简约而又特别。现代的元素与强烈的时尚感并存，流露出一种新一代都市群体对待生活的态度。

This case is based on modern black and white fashion style, and designers choose cream-colored oak as a toner of the space to create a modern, simple, fashionable, individualized space with a feeling of coziness and elegance.

On the basis of modern simplicity style, designers add house owners' likes and personality in the living room, dining room, children's room and main bedroom. The black-white urban city photo in the dining room displays personality and art. Hanging decorations on the bedside represent a sense of humor, very simple but special. Modern factors coexist with fashion, indicating the new urban generations' attitude towards life.

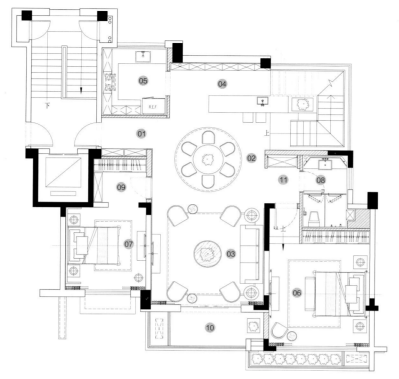

01	Porch
02	Dining Room
03	Living Room
04	Water Bar
05	Kitchen
06	Bed Room
07	Bed Room
08	Bath Room
09	Cloak Room
10	Balcony
11	The Hall
01	入户玄关
02	餐厅
03	客厅
04	水吧区
05	厨房
06	卧室一
07	卧室二
08	卫生间
09	衣帽间
10	休闲阳台
11	过厅

First Floor Plan
一层平面布局图

01	Porch
02	Library
03	Entertainment Terrace
04	Master Bedroom
05	Second Bedroom
06	Cloakroom
07	Cloakroom
08	Main Bathroom
09	Family Bathroom
01	内玄关
02	书房
03	休闲露台
04	主卧室
05	次卧室
06	衣帽间
07	衣帽间
08	主卫
09	公卫

Second Floor Plan
二层平面布局图

345

济南路劲F户型样板房

济南 / 中国

此户型方案用现代简洁的手法表现出低调奢华的都市情怀。设计师在材质和色彩的搭配上，十足的考究。整个空间以光漆面的金箔木饰面搭配啡色皮硬包并配以精致的不锈钢条与做旧银箔、马赛克等处理，形成干净而凝练的表达，有力却不失细节和质感。同色系、同纹理的组合，为冰冷的空间注入温暖的语言，散发着现代都市新贵尊贵的气质。

JI'NAN ROAD KING F-TYPE SAMPLE HOUSE

JI'NAN / CHINA

The case expresses urban low-key luxury by simple modern design method. Designers are very particular about the match of materials and colors. Material arrangement like gold foil wood veneer covered with lac varnish, brown leather, delicate stainless steel, faded silver foil, mosaic etc., is very clean and complete, showing both strength and sense of quality and details. Combination of tone on tone and fixture on fixture injects warm feeling into the cold space, indicating the unique taste of modern urban elites.

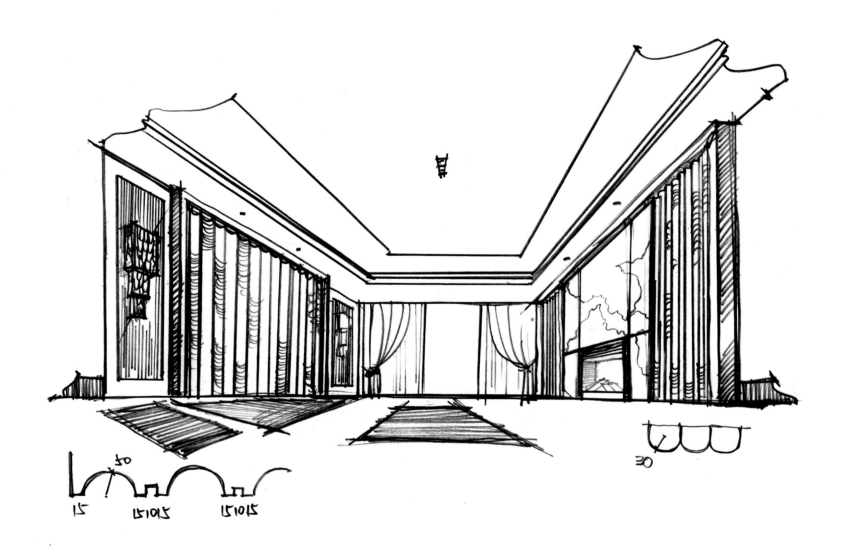

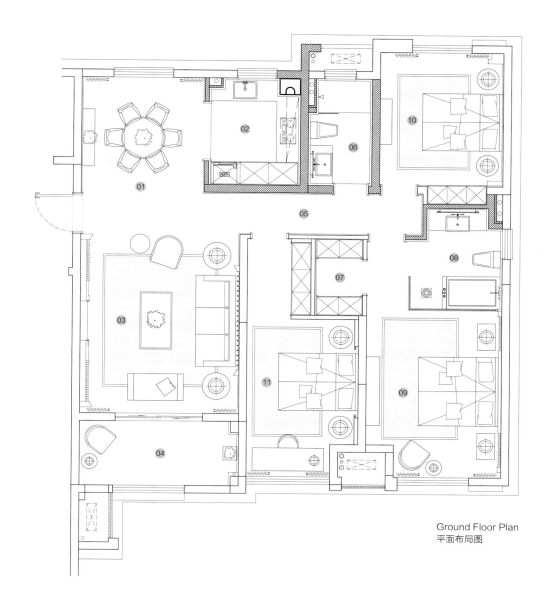

01	Dining Room	01	餐厅
02	Kitchen	02	厨房
03	Living Room	03	客厅
04	Balcony	04	阳台
05	Hallway	05	过厅
06	Family Bathroom	06	公卫
07	Cloakroom	07	衣帽间
08	Main Bathroom	08	主卫
09	Master Bedroom	09	主卧室
10	Second Bedroom	10	次卧室
11	Children's Room	11	小孩房

Ground Floor Plan
平面布局图

合肥万科森林公园17栋一层样板房

合肥 / 中国

在整个温暖的色调中，设计师将东方的元素以现代的手法融入整个空间。纯粹的科技木饰面配以金色的不锈钢条，精致的贝壳马赛克点缀其中，再与浅色麻面墙布搭配，使整个风格既保留了现代都市的精致奢华又不失东方情调的意境与沉稳。正所谓"进可守得住繁华，退可享受到宁静"。

HEFEI VANKE FOREST PARK BUILDING 17 ONE-FLOOR SHOW FLAT

HEFEI / CHINA

Oriental factors are mixed into the entire space of warm color. Pure technological wood veneer and golden stainless bar adorned with delicate shell mosaic, matching with light linen wall paper, kept both the delicate luxury of modern city and the solemnity and sense of Oriental style. That is what we say "One step forward we see prosperity; one step backward we enjoy serenity".

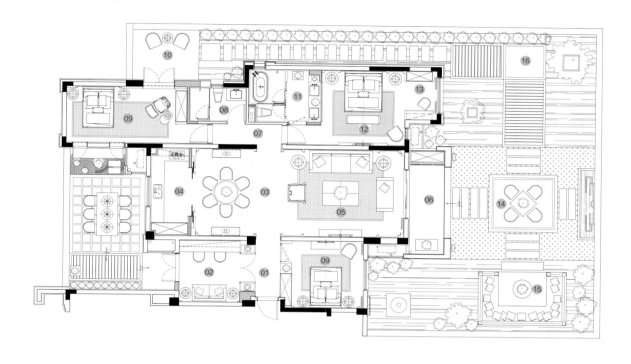

Ground Floor Plan
平面布局图

01	Entrance Hall	01	玄关
02	Multifunction Room	02	多功能房
03	Dining Room	03	餐厅
04	Kitchen	04	厨房
05	Living Room	05	客厅
06	Landscape Balcony	06	景观阳台
07	Hallway	07	过厅
08	Second Bathroom	08	次卫
09	Second Bedroom	09	次卧室
10	Landscape Lounge	10	景观休闲区
11	Main bathroom	11	主卫
12	Master Bedroom	12	主卧
13	Dressing & Powder Room	13	更衣化妆间
14	First Floor South Courtyard Lounge	14	一层南院休闲区
15	Landscape Relaxing Area	15	四季亭
16	Landscape Lounge	16	景观休闲区

合肥万科森林公园17栋二层样板房

合肥 / 中国

HEFEI VANKE FOREST PARK BUILDING 17 TWO-FLOOR SHOW FLAT

HEFEI / CHINA

此户型销售期间主要面向城市精英阶层,故设计师的设计意图很清晰,就是将低调奢华的空间,用现代简洁的手法表现出来。尼斯木立体的纹理搭配黑镜钢的力量,再配以质感强烈的皮纹以及绒布软包,低调干练,传达精致的感情。软装上配以新古典的简约家私,或皮或绒的面料干净而凝练的表达,有力却不失细节和质感。同色系、同纹理的组合,为冰冷的空间注入温暖的语言,散发现代都市精英新贵的独特气质。

The house is designed for urban elites, so the design style is clearly featured of low-key luxury presented by simple modern design method. Three dimensional fixture of Lacewood matches with the strength of black mirror steel and quality sense of leather and flannel, expressing simple and delicate feelings. For soft decorations, designers choose new classical furniture, clean and tidy leather or flannel, showing both strength and sense of quality and details. Combination of tone on tone and fixture on fixture injects warm feeling into the cold space, showing the unique taste of modern urban elites.

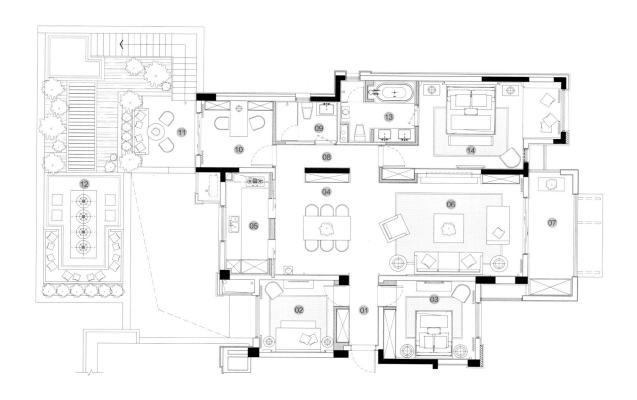

Ground Floor Plan
平面布局图

01	Entrance Hall	01	玄关
02	Multifunction Room	02	多功能房
03	Second Bedroom	03	次卧室
04	Dining Room	04	餐厅
05	Kitchen	05	厨房
06	Living Room	06	客厅
07	Landscape Balcony	07	景观阳台
08	Hallway	08	过厅
09	Second Bathroom	09	次卫
10	Study	10	书房
11	Terrace	11	露台
12	Family-Gather Area	12	家庭聚会区
13	Main Bathroom	13	主卫
14	Master Bedroom	14	主卧

TRANSCRIPT OF ANDREW MARTIN INTERVIEWING MATRIX DESIGN

ANDREW MARTIN 对矩阵纵横进行采访文稿

Most interior design agencies are of a simple organizational structure: founder, design brains and managers. Such structure goes with the characteristics of design industry; however, when a company develops to certain stage, it will constrain the continuing expansion of the company. Matrix Design always presents as a team, whose new company organizational model makes it a special group in design field. Mathematically speaking, matrix is arranged by many numbers, horizontally and vertically, to form rows and columns, calculating according to Higher Algebra to find the answer. Matrix Design is such a matrix, among which each element plays an important role. Some elements are good at hospitality and club design, some are specialized in business show house design, and others have special management skills. Each element in the team is arranged criss-crossly to generate energy and productivity together so as to achieve Matrix's design works.

The works created by Matrix Design are conform to high standard, thus been selected into "ANDREW MARTIN International Interior Design Awards Yearbook" known as "The Interior Design Bible" for three times. Neo-Asia style becomes the design label for Matrix Design. Besides unique design techniques and design management, special design thoughts are also one of the reasons why these new design talents are widely recognized.

Unmixme Wang, the Founder, defines Matrix's design techniques as designing in a mathematic way. The group cooperation is based on the module analysis on projects, the calculation of structural reformation, the partition and decoration of hexahedron, as well as the formulization and theorization of perceptual design tasks. The most distinct DNA of the Matrix's young group is Internet design thoughts. As most post 80s individuals, the team is also born in the Internet era, therefore deeply influenced by Internet. The uniqueness of Internet is the planar communication among people, which is more direct and open without face to face contact. The founders, Unmixme Wang, Idmen Liu and Zhaobao Wang, met on IDMEN design forum. They trust each another and come together after frank communication. As Unmixme Wang said, Matrix Design succeeded as a team due to "trust", just similar to the core idea of "Naruto" — the comic he likes most. They trust the power of "trust". Even though the word "trust" is simple, yet it is hard to reach for business partners.

多数室内设计公司的组织架构都极为简单，创始人兼任设计主脑和管理者。简单的组织架构虽然很符合设计行业的特点，但发展到一定阶段，会制约公司的持续壮大。从来都是以一个团队出现的矩阵纵横，因为全新的公司组织模式而成为设计行业极为特殊的一个组合。在数学中，矩阵由一个个数字元素组成，数字纵横排列，构成行和列，矩阵按照高等代数的规则进行运算，得到答案。矩阵纵横团队就是一个矩阵，其中每一个"元素"都各有专长。有的擅长酒店会所规划，有的擅长商业样板房设计，有的擅长统筹管理。团队的每个"元素"被纵横排列，集合迸发出能量和效率，给出矩阵纵横的设计答案。

矩阵纵横的作品具有很高的设计水准，先后三次入选素有"室内设计圣经"之称的《ANDREW MARTIN国际室内设计大奖年鉴》，新亚洲风格已成为代表矩阵纵横的设计标签。解读这群设计新锐能够被认可的原因，除了独特的设计方法和设计管理，他们还拥有独特的设计思维。

创始人王冠把矩阵纵横的设计方法定义为用数学的方式做设计。创作作品的模块分析、结构改造的加法减法、空间六面体的分拆装饰，把感性的设计工作公式化、理论化是团队的协作基础。互联网设计思维是矩阵纵横这个年轻的团队最独特的DNA。和很多80后一样，这个团队诞生于互联网时代，同时被互联网深深地影响着。互联网的特殊性在于人与人的平面式交流，因为不是面对面所以沟通直接，毫无保留。合伙创始人王冠、刘建辉、王兆宝在设计论坛IDMEN上结识，坦诚的交流积累了信任，信任让他们走到了一起。王冠说矩阵纵横之所以能以团队的形式成功，根本原因和自己喜欢的漫画《火影忍者》的核心思想一样，即两个字——"相信"。他们相信"相信"的力量。这两个字看似简单，对于商业合作者来说其实很难。

刘建辉除了矩阵纵横合伙创始人身份外，还是《IDMEN室内中国网》的创办者。上网泡论坛似乎不是讲求艺术体验的设计师应有的生活，但刘建辉说恰恰是这样的经历让他完成了很多对室内设计的艺术积累。年轻设计师在求学阶段不会有很多的机会或者足够的经济实力去体验好的设计，书本和互联网就变成了最简单最便捷的设计体验形式。年轻的刘建辉以互联网为起点，完成了从学习到理解，从理解到创新，进而形成自我风格的成长过程。凭借对设计的狂热和敏锐的设计本能，他一直在对如何将建筑、室内设计、平面视觉以及家具、灯具和配饰设计融为一体进行有效思考和探索。在每个项目中，他都鼓励自己及设计团队能够实现甲方业主要求的前提下兼顾创新性、感官吸引力和实用效果。通过坚守本土文化与全球视角相互融会贯通，他总能用有限的空间感观去创造无限的遐想；并通过光线、色彩、质地以及体现适时的设计语汇来展现自己的设计。

王兆宝是矩阵纵横的大管家。设计师出身的他，用设计思维做项目管理的同时也用互联网思维管理设计。在创办矩阵纵横之前，王兆宝有过一次因甲方工程撤资而项目搁浅的经历，这让他半年的付出徒劳无功。创办公司后，作为管理者，他希望自己的经历不再重演，设计师所付出的劳动不必为非设计原因买单，于是独创了"屠龙积分"式的设计管理模式。曾经是网游公会头目的王兆宝深谙游戏中对公会成员的管理和对财富的分配，并借鉴到公司的管理实践中。他给每个设计项目设定总积分，团队成员按设计时间和贡献获得各自的积分，最终这些积分可以根据整个公司的产值按比例兑换成薪酬。互联网生活影响着新锐设计团队的思想，也催生出新的设计解题方式。

王冠、刘建辉、王兆宝只是矩阵中的三个"数字元素"，矩阵中还有，而且将会有更多的"数学元素"和他们一起，排列成阵，纵横天下。

采访时间：2014年11月20日

英国ANDREW MARTIN 中国区负责人 卢丛周先生

Idmen Liu is not only one of the founders of Matrix Design, but also the founder of IDMEN.cn. Joining an online forum shouldn't be the life of a designer who is particular in artistic experience. However, through this platform he obtained rich accumulation of interior design art. Young designers have less access and limited money to experience fine design during their studying stage. But books and Internet become the simplest and fastest access to experience design works. The young Liu started from Internet, and completed the personal style development process from learning to comprehending then to creating. With enthusiasm and sharp design instinct, he thought and researched to blend architecture, interior design, plan design, lighting and other accessories together. In each project, he encourages himself and his team to achieve what clients expect, and to achieve fully considerate creativeness, sensual appealing and practical usage. He can create infinite imagination with finite spatial sensation through the combination of local culture and international perspectives. He distinguishes his works by light, color, texture and appropriate design languages.

Zhaobao Wang is the majordomo of Matrix Design. Started as a designer, he manages projects with a design thought and manages design with an Internet thought. Before Matrix Design was formed, he had a failed project because client withdrew the fund, which made a half year efforts go in vain. Hence, after the company established, he, as a manager, does not want to go over such experience again. Designers' efforts should not be wasted for off-design reasons. Therefore, he created a design management mode — "Dragon Kill Point". As an experienced online game guild leader, he was familiar with the member management and fortune distribution rules in the games, so borrowed it to management practices. He assigns each design project certain points, and the team members acquire their points according to the design cycle and contribution; finally, these accumulated points can exchange salary proportionately according to the economic output. Internet life influences the thoughts of new design teams, helps to generate new design solutions.

Unmixme Wang, Idmen Liu and Zhaobao Wang are just three "Mathematical Elements" among the matrix which has more elements together to form these arrays and to lead the world.

Interview Date: November 20, 2014
Ricky LU, Head of Greater China region,
British Andrew Martin Awards

ABOUT A BOY
关于一个男孩

ZHONG JIAN / DESIGNER OF MATRIX　钟坚 / 矩阵纵横设计师

"I'm back again!" His Wechat blog revived after he was able to type with a touchpen in mouth, he is much more optimistic than me though we are both Scorpios, but in my mind his image is still in that dark late autumn.

His net name was "poor kid" before 5th Nov. 2012. This hapless name! I'll never forget 10:12 am that day, he was on bed in normal times, but that day he abnormally posted a Wechat blog saying "Love whom you want to love, see whom you want to see, hug your best love though cold in late night when you still can; because regret is not fearful, what really fearful is forever lost, losing every possible life." This abnormal message became a starter of the nightmare. At 20:17, he fell down on the badminton field after a message "Haven't played for a long time". I used to sleep early but I happened to have not turned off my phone that night, I was informed at 21:00 and only saw him twitching on emergency sickbed with foaming in mouth when I got to hospital. The emergency physician said it might be epilepsy, he would wake up soon after medicine treatment. Two colleagues accompanied him through the night; we were all waiting for him to wake up.

Finally the morning time next day, professional doctors came to work. A feminine doctor holding his X-ray film announced his death sentence to another doctor, "This boy is done, focus entered into brain stem." Everyone present was stunned on this news, such a strong man turned

"我又回来了！"同属天蝎座的他内心比我阳光多了，自从他可以用口叼着触笔打字后他的朋友圈就复活了，可是在我内心他始终定格在那个黑暗的深秋。

在2012年11月5日之前，他的网名一直是"可怜的孩子"，这个倒霉的名字！我永远不会忘记那天上午10点12分，他反常地发送了一条微信。这个点平时他还在被窝！"趁着能的时候，去爱你想爱的人吧，去见你想见的人吧，哪怕深夜寒冷去拥抱你的最爱吧，因为遗憾不可怕，可怕是永远的失去，失去一切可能的生命。"这条反常的信息成了一切噩梦的开始，在当天20点17分发送了一条"很久没打了"的信息后，他就瘫倒在了羽毛球场上。习惯早睡的我那天恰好还没有关机，21点接到电话赶到医院的时候看到他口吐白沫地在急救病床上抽搐。急诊医生说可能是癫痫，用了药他一会就能醒过来，两个同事陪着他过夜，这一整晚我们都在等着他醒过来。

终于在第二天的上午，当专职医生们来上班的时候，一位女医生拿着他的片子对着另一位医生宣判了他的死刑："这孩子废了，病灶进入脑干。"在场的所有人都傻眼了，平时身体健壮的他居然在一夜间变成了植物人！意识模糊，肺部感染，插着各种管子，只有在剧烈的咳嗽时身体才能本能地在病床上弹起，除此之外就只有眼珠可以转动，他的病症最后被确诊为"先天性血管畸形引起的脑血管病变"，又称"闭锁综合征"。

"闭锁综合征"相关资料

患者大脑半球和脑干被盖部网状激活系统无损害，因此意识保持清醒，对语言的理解无障碍，由于其动眼神经与滑车神经的功能保留，故能以眼球上下示意与周围的环境建立联系。但因脑桥基底部损害，双侧皮质脑干束与皮质脊髓束均被阻断，外展神经核以下运动性传出功能丧失，患者表现为不能讲话，有眼球水平运动障碍，双侧面瘫，舌、咽及构音、吞咽运动均有障碍，不能转颈耸肩，四肢全瘫，可有双侧病理反射。因此虽然意识清楚，但因身体不能动，不能言语，常被误认为昏迷。脑电图正常或轻度慢波有助于和真正的意识障碍相区别。

into a "vegetative being" overnight! Clouding of consciousness, pulmonary infection, various tubes all over his body. He hardly had any moves except for rolling eyeballs and bouncing up from sickbed for intense coughing. At last he was diagnosed as "Congenital vascular malformation caused by cerebrovascular disease", also known as "locked-in syndrome".

DATA ABOUT "LOCKED-IN SYNDROME"

Locked-in syndrome (LIS) is a condition in which patients are aware but cannot move or communicate verbally due to complete paralysis of nearly all voluntary muscles in the body except for the eyes. There is no damage on patients' cerebral hemisphere and brain stem of the reticular activating system, so they have consciousness. Locked-in syndrome usually results in quadriplegia and the inability to speak in otherwise cognitively intact individuals. Those with locked-in syndrome may be able to communicate with others through coded messages by blinking or moving their eyes, which are often not affected by the paralysis. Patients who have locked-in syndrome are conscious and aware, with no loss of cognitive function. They can sometimes retain proprioception and sensation throughout their bodies. Some patients may have the ability to move certain facial muscles, and most often some or all of the extraocular eye muscles. Individuals with the syndrome lack coordination between breathing and voice. This restricts them from producing voluntary sounds, though the vocal cords are not paralysed. Unlike persistent vegetative state, in which the upper portions of the brain are damaged and the lower portions are spared, locked-in syndrome is caused by damage to specific portions of the lower brain and brainstem, with no damage to the upper brain.

Clinical features:

1. Conscious clear, able to understand verbal talking, able to understand questions and can answer by blinking eyes or rolling eyeballs.
2. Limbs paraplegia, bilateral pathological reflex being positive.
3. Aware of pain stimulus and sound, normal on audition, hemidysesthesia sometimes, decerebrate rigidity at body stimulation.
4. With poor prognosis. Patients usually die within hours or days.

Patients able to survive for days are rare. Neither a standard treatment nor a cure is available. Stimulation of muscle reflexes with electrodes (NMES) has been known to help patients regain some muscle function. Other courses of treatment are often symptomatic. Assistive computer interface technologies, such as Dasher, combined with eye tracking, may be used to help patients communicate,

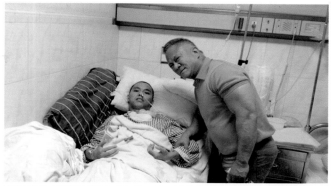

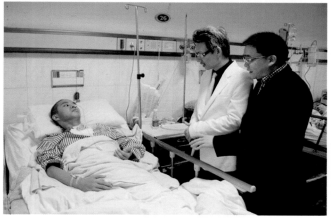

临床表现：

1. 意识清楚，能听懂别人讲话，明白问话，可用睁、闭眼或眼球活动示意回答。
2. 四肢全瘫，双侧病理反射阳性。
3. 对疼痛刺激及声音能感知，听力正常，偶有偏身感觉障碍，刺激肢体可出现去脑强直。
4. 预后差，多在数小时或数日内死亡，能存活数日者少见。

闭锁综合征没有标准的治疗方法，也没有可以完全治愈的方法。电极刺激肌肉反射可用来帮助患者恢复一些肌肉功能，其他的治疗方法往往都只是对症治疗。辅助计算机界面技术通过对患者眼球活动的追踪可以用来帮助患者和外人沟通。闭锁综合征预后差，患者运动功能显著恢复的情况非常罕见。大多数闭锁综合征患者运动控制无法恢复，但可以通过仪器设备帮助患者进行沟通。在闭锁综合征发病后的前四个月，患者病死率高达90%。但也有少数人可以生存很长一段时间。

在接下来几个月的时间里我们动用了所有可能，企图与他的命运对抗，在网络上不停地发布，寻找可能对此有用的信息，但是最后的结果依然是残酷无情的，这种病无法医治，只能慢慢地康复。说是植物人其实也是不正确的，因为他的意识是完全清醒的！在我看来，这才是最最残忍的地方，每天只能游魂般地面对这个世界……

就这样两年多的时间过去了，我不知道他是如何坚持过来的，每当想到在病床上的他，我都无法控制自己的情绪，听到他喜欢的歌曲

Extremely rarely does any significant motor function return. The majority of locked-in syndrome patients do not regain motor control, but devices are available to help patients communicate. Within the first four months after its onset, 90% of those with this condition die. However, some people with the condition continue to live much longer.

In the following months, we tried every possibility to help him fight with destiny by consulting doctors and searching online to find every possible useful information; however, the result is cruel and merciless—this disease can not be cured, only has chance of slowly healing. "Vegetative being" actually is not appropriate for his situation because he has very clear consciousness! In my eyes, that is the most cruel suffering, soberly facing the world like a wandering soul...

Two years have passed, I don't know how does he make through all the time. Every time I think of him on sickbed I lost control of my emotion, every time I hear his favorite songs I shed floods of tears. I tell jokes to him on every visit, he smiles happily, but the more cheery his smile the more wrenched my heart. My heart broken, for unfairness of God and helplessness of what we can do! Yes, we are both Scorpios, yet he is much more optimistic than I am, much stronger than I am though his parents divorced in his childhood, he worked so hard and just bought a new house in his hometown before the illness. He said he would accompany his mother spending her remaining years in comfort...

What a poor kid! Why did you choose for yourself such a hapless net name! But he is lucky in the meantime, God bless him there were colleagues by his side when the sickness attacked; the consequences would have been unimaginable if he were alone at that time! Lots of warm-hearted net friends, bosses in design field, industrial leaders and colleagues came to visit and donate for him, and the brothers and sisters in our company, we undertook all the fee for cure and recovery, all of us stand with him for accompany and support. Now I understand why his smiles so bright, because he knows very well it is the most precious feedback he gives to us.

Yes, the story has not ended yet, we will accompany and support him in future days like we always did. We arranged computer and Internet to encourage him communicate with the outside like I wrote in the beginning. I hope one day he can write his own legend through his efforts! His name is—Zhong Jian, from Pingxiang Jiangxi, 28 this year!

2014.12.21 noon Unmixme Wang

都会泪如雨下，虽然每次去看望他都跟他讲着各种笑话。他开心得笑着，他越是笑得灿烂我越是心如刀绞。为上天的不公，也为我们的无能为力而感到心碎！是的，同属天蝎座的他内心一定比我阳光，比我强大，从小父母离婚，拼着命努力奋斗，就在病倒之前还在老家买了房子，他说要把妈妈接过来安度晚年……

哎，可怜的孩子！你为什么给自己取了这么一个倒霉的网名！但他确实也是幸运的，上帝保佑他发病时有同事在身边，如果当时只身一人后果将不堪设想！病倒后有大批的好心网友、设计大佬、业界领导和同仁们都纷纷前来探望捐款，还有公司的兄弟姐妹，大家都不离不弃地陪伴着他，并负担了他所有的治疗及康复的费用。现在我能明白为什么他的笑容那么灿烂阳光了，因为他知道这是他能给予我们最珍贵的回报。

是的，故事还没有结束，在接下来的岁月中我们将会一如既往地陪伴他，并且给他配置了电脑和网络，鼓励他与外界交流，就像一开始写到的那样。我希望有那么一天他能通过自己的努力书写属于他自己的传奇人生！他的名字叫 —— 钟坚，江西萍乡人，今年28岁！

2014.12.21午 王冠

POSTSCRIPT

Zhong Jian, net name "Monk Designer", was very famous on the top designer websites for his excellent design sketches. He joined in Matrix in 2010 and has won several well-known design awards since then.

Zhong Jian is now accepting rehabilitation in Nan'ao People's Hospital and Shenzhen Second People's Hospital. He needs to change from one hospital to the other every 2 months and has been doing this for over 2 years; the difficulties are unimaginable. He is allowed to spend one hour online everyday— his mother holds the Ipad for him from 8:00—9:00 pm. Sometimes he chats with us, but since he is still unable to move his body, he types every single letter with a pen in his mouth.

However, the cheering news is that his body has been all the time going on slow recovery, and he is all the time fighting with the disease. The company has supported him for all his medical fee and will do this all the time, but this is a long process, so we'd like to provide a donating account if anyone wants to help him a little. And it will be nice to contact him on Wechat if you want to give him some encouragement and comfort.

Wechat: 258128569 (Online time: 8:00 — 9:00 pm)
Donating bank account:
Shenzhen Rural Commercial Bank Nan'ao Sub-branch
6230 3518 1640 1869   Li Runan (Zhong Jian's mother)

后记

钟坚网名为"和尚也设计",当时凭借高超的效果图技术曾风靡各大设计网站,在各大设计类网站有很高的知名度,2010年加入矩阵,先后获得多项知名设计奖项。

现在钟坚在南澳人民医院和深圳市第二人民医院两所医院进行康复,每隔2个月就需要更换一次地方,到现在为止已经2年多了,其中的艰辛可想而知。他母亲基本每天晚上8—9点时会拿着IPad让他上一个小时的网,他有时候会跟我们聊天,每一个字都是用嘴叼笔一个字母一个字母地点上去。他身体还是不能动弹,但值得可喜的是他一直都在慢慢恢复,一直在跟病魔抗衡。到现在为止钟坚的所有医药费用都还是公司在资助,但是这是个漫长的过程。如果大家也想为他出一份力量,我们会提供一个捐款账号,如果大家想给他一个鼓励、一份支持,可以联系他的微信。

微信:258128569(在线时间:晚上8—9点)
银行账户:深圳农村商业银行南澳支行
6230 3518 1640 1869 李汝南(钟坚母亲)